Praise for

Essential Light Straw Clay Construction

Light straw-clay is one of the most versatile natural wall systems around, combining straw's insulation value with clay's thermal mass and preservative qualities for excellent performance in almost any climate. Why is it so little known? Partly due to a shortage of reliable how-to information in English. Lydia Doleman has solved that problem once and for all with this brilliantly written and illustrated guidebook. Absolutely the most useful book on the topic.

— Michael G. Smith, co-editor, *The Art of Natural Building* and co-author, *The Hand-Sculpted House*

Lydia Doleman is the best, and this book is the best. It is the book that Lydia would have liked to have had at the start of her journey to become one of the most talented, experienced, and professional natural builders in North America. Instead, after half a lifetime of designing and building and building and building, she has gathered it all together for us to have and use. This book really is for all of us, so we can go forth and use our own hands to build a better, more healthful and beautiful world.

— Mark Lakeman, Founder and sustainer of communitecture and City Repair

Lydia Doleman has written a "must have" resource on the multiple techniques for Light Straw Clay construction. Well researched and grounded in her own 20 years of hands-on natural building experience, her love of the craft, and her attention to detail shine through in every page.

— Robert Laporte and Paula Baker-Laporte, authors, *The EcoNest Home*

Essential Light Straw Clay Construction gives building professionals and owner-builders the knowledge and an array of options to design and construct a durable, resource-efficient, and high-performance home with the simple materials of earth, straw, and wood.

— Martin Hammer, architect, co-author of the *Light Straw-Clay* appendix in the International Residential Code

For thousands of years, straw and clay formed the bedrock of human habitat. The natural building revolution brought us many books on strawbale, cob, rammed earth, earthbags, hempcrete and adobe, but until now almost nothing on my favorite recommended style for ecovillages in every climate — light clay straw. Thanks to Lydia Doleman for the giving the world this comprehensive construction companion.

— Albert Bates, author, *The Post-Petroleum Survival Guide*, *The Biochar Solution*, and *The Paris Agreement*

We have been teaching natural building techniques since constructing our first Honey House earthbag dome in 1996. Yet in 2014 we chose the Light Straw Clay system to build a 300 square foot kitchen addition onto our existing stick frame house. We are living proof that Lydia's prescriptions and precautions will prepare you for a positively playful, pleasantly proficient workplace designed to produce a prize winning project. All presented in Ms Doleman's heartfelt pragmatic pioneering spirit that personifies the joy of natural building. Thank you Lydia, for providing a neat package of protocols that promote the virtues of light straw clay into our modern day perception.

— Kaki Hunter and Doni Kiffmeyer, authors, *Earthbag Building*

Having constructed several strawbale homes I found this book very informative comparing the similarities and differences to light straw clay.

— Matt Bostwick

sustainable
building
essentials

essential

LIGHT STRAW CLAY CONSTRUCTION

the complete step-by-step guide

Lydia Doleman

new society
PUBLISHERS
www.newsociety.com

New Society Sustainable Building Essentials Series

Series editors
Chris Magwood and Jen Feigin

Title list

Essential Hempcrete Construction, Chris Magwood
Essential Prefab Straw Bale Construction, Chris Magwood
Essential Building Science, Jacob Deva Racusin

See www.newsociety.com/SBES for a complete list of new and forthcoming series titles.

THE SUSTAINABLE BUILDING ESSENTIALS SERIES covers the full range of natural and green building techniques with a focus on sustainable materials and methods and code compliance. Firmly rooted in sound building science and drawing on decades of experience, these large-format, highly illustrated manuals deliver comprehensive, practical guidance from leading experts using a well-organized step-by-step approach. Whether your interest is foundations, walls, insulation, mechanical systems or final finishes, these unique books present the essential information on each topic including:

- Material specifications, testing and building code references
- Plan drawings for all common applications
- Tool lists and complete installation instructions
- Finishing, maintenance and renovation techniques
- Budgeting and labor estimates
- Additional resources

Written by the world's leading sustainable builders, designers and engineers, these succinct, user-friendly handbooks are indispensable tools for any project where accurate and reliable information is key to success. GET THE ESSENTIALS!

Cover design by Diane McIntosh. Cover images ©Lydia Doleman, except shed ©Rob West
Interior Illustrations by Dale Brownson.
Thumbs up art: AdobeStock_23490949.
Straw bale chapter photo: Adobestock_94385282. Sidebar photo AdobeStock_49425959.

Printed in Canada. First printing May 2017.

Any other inquiries can be directed by mail to:
New Society Publishers
P.O. Box 189, Gabriola Island, BC V0R 1X0, Canada
(250) 247-9737

Library and Archives Canada Cataloguing in Publication

Doleman, Lydia, 1976-, author
Essential light straw clay construction : the complete step by step guide / Lydia Doleman.

(Sustainable building essentials)
Includes bibliographical references and index.
Issued in print and electronic formats.
ISBN 978-0-86571-843-2 (softcover).--ISBN 978-1-55092-638-5 (PDF).--
ISBN 978-1-77142-232-1 (EPUB)

1. Building materials--Environmental aspects. 2. Dwellings--Design and construction. 3. Ecological houses--Design and construction. 4. Sustainable buildings--Design and construction. 5. Straw. 6. Clay.
I. Title. II. Title: Light straw clay construction. III. Series: Sustainable building essentials

TH4818.S77D65 2017 693'.997 C2017-901525-7
C2017-901526-5

Funded by the Government of Canada | Financé par le gouvernement du Canada | Canada

New Society Publishers' mission is to publish books that contribute in fundamental ways to building an ecologically sustainable and just society, and to do so with the least possible impact on the environment, in a manner that models this vision.

FSC MIX Paper from responsible sources FSC® C016245 www.fsc.org

Certified B Corporation

new society PUBLISHERS www.newsociety.com

Contents

Acknowledgments

A SPECIAL THANK YOU TO THE MANY PEOPLE who have inspired me through my building career! Too many to name, but a very special thank you to Ann Beninger for shining a light on the path before me way back in 1997, thanks to you I saw the straw, the wood, and the clay, and I have never looked back!

An extra special thank you to those who have encouraged me on this epic journey of writing this book. Thank you Brad Lancaster for reminding me to write about what is "juicy for me," and Tyler Davis for giving me reality checks and the numbers of publishers, and Michael G. Smith for really making me see it was possible and being up for the adventure, and Sukita Crimmel and James Thomson for leading the way, and Clay Glad for editing the first renditions, and Robert and Paula Laporte for cleaving the path for LSC to blossom in the States. And thank you to the many people who contributed photos: Jim Rieland, Jared Smith, Mark Lakeman, Michael G. Smith, Erica Ann Bush, and the Laportes, and a special thank you to Chris Magwood and Jen Feigin for editing this series, handling my questions and helping push this baby into the world.

An extra special thank you to Martin Hammer for his commitment to the unique challenge of creating the codes for natural wall systems enabling so many people to utilize these methods and for his valuable help in improving the information in this book (especially Chapter 11).

And a super-duper thank you to my family and family of friends who have been there through many projects, and many dreams, rain or shine. Thank you Dad ("Wanna see me jump?" aka: Look! I wrote a book!!) and Devery for being such a part of who I am today.

Thank you Matt Musselwhite for parenting with me and teaching me about strength and freedom. Thank you Matt Vogel for your endless compassion and deep connection. Thank you Caroline Musselwhite for being a relentless writer and supporting me 100% in this project and thank you Maud Powell for being such a dear friend and continual source of inspiration. I am honored to have you all in my life!

This book is dedicated to my mothers, Madeline Doleman and Donna Doleman, both of whom loved me no matter how much dirt I tracked into the house, supported me on my unique life path, and left this planet way too early; and to my daughter Madeline Siskiyou, may the world hold you with the warmth and safety of a handmade home.

Chapter 1

Introduction

EARTH: OUR HOME OF HOMES. For thousands of generations it has also been our building material of choice: raw, local, and reflective of people, place, and a logical human scale. Extravagance was saved for structures and buildings with profound cultural significance: large timbers for a sacred house in a scrub brush desert; stones the size of school buses dragged 12 miles and up a mountain to build a temple; miles and miles of monolithic earthen walls to keep out invaders. Extreme structures like these required a concentration of human population and centralized power to make them happen. The vast majority of human dwellings have seamlessly ebbed and flowed from and back into the landscape over the centuries. In the last 200 years we have seen a radical transformation of human habitation, in location, building size, and material choices. Even in this era of industrialized and commodified building systems, it has been estimated that over 50% of the earth's population still build with and reside in houses made of earth. (Ronald Real, *Earth Architecture*, 2009). From the adobe pueblos of the North American Southwest to the 13-story cob towers of Yemen, people around the world inhabit earthen structures — and they've been doing so for thousands of years.

Most people living in the Western World live in cities; many more will soon move to sprawling cities; and millions are experiencing the urbanization of rural areas. Most North American homes are relatively recently constructed; the materials used to build them came from the lumber yard, hardware store, or giant box store. Only a few generations ago, these homes would have been built with indigenous materials from the forest, field, or directly under foot. The curse of Western civilization is that, despite its vast quantity of available conveniences, it fosters a profound and insidious culture of disconnection. The commercialization of all needs has created a social architecture that divorces people from place, from planet, and from each other.

In the last 50 years there has been a resurgence of interest in rehabilitating severed connections on multiple levels — from social reconnection (charity, co-housing, neighborhood involvement, community-creating) to material reconnection with the items of daily life (where food and fuel come from, where wastes go, where building materials are sourced, and how all of the above is transported and transformed from raw materials).

The need for these types of connections drives a culture on a conscious and subconscious level. The empowerment and purpose someone feels when they feel and understand the history and meaning of the materials of their daily life is profoundly different from the probable dullness and hollowness of someone's experience who does not. In Polynesia, the construction of a dwelling involves the entire community. All materials are sourced locally and are assembled with songs and meaning assigned to each facet of the process and each part of the building. In essence, the whole home resonates with symbol and meaning. In the Western World, the essence of the home is its value as an investment — a commodity described in cost-per-square-foot terms; songs ascribed to any facet of these buildings are likely to be the curses and frustrations expressed by those who have built them more for money than any love of any "craft."

For those who are longing for spiritual connection, to experience a deeper sense of the sacred, or — on the most basic level — begging to feel some sense of meaning in their lives, one positive action to take is starting a collection of things made by hand by people they know. A carved spoon, a clay bowl, a knitted scarf: all these items carry with them a significance that cannot be purchased. Consciously seeking to incorporate items with significant stories into one's life can be a powerful experience. They become landmarks of connection throughout a routine existence. This might sound like a mundane exercise, but just as small reminders in daily life can horribly trigger a sufferer of PTSD, these objects with history and meaning can be the positive opposite.

In this rising tide of connection and resiliency, certain notions and movements have been playing major and convergent roles. The local and organic food movement is reconnecting people with food and sustainable systems of production. The environmental movement is reconnecting people with our place in the natural world in a holistic and reparative way. The social justice movement is reconnecting people with their own power and with each other. The sustainability movement is connecting and educating people with the multifaceted knowledge required to inhabit and sustain one's self and greater community in a particular place and climate. For many, this is known as *permaculture.* A major part of permaculture is *whole-systems design.* This is looking at the design of a garden, home, community, and or other system through a long-term lens that takes all factors into consideration: location, geology, climate, soil type, aspect, natural resources, people, etc. The facet of a whole-systems design process that takes shelter into consideration (whether you are a permaculturalist or not) is most commonly referred to as *natural building.*

Natural building, in its most basic definition, is a method of construction that uses minimally processed and locally sourced "natural" materials. In its most embodied form, it represents a whole-systems approach to construction that incorporates a worldview into a structure. It wholeheartedly takes into consideration the grand global implications and footprint (in every sense of the word) of materials, including acquisition, processing, transport, intended use, and post-use. For some, it's just an earth-friendly and breathable wall system, for some, it's the structural implementation of permaculture; and for others it is a means to healthy and nontoxic living or a method to exit the industrialized capitalist banking system. For whatever reason it draws people, natural building as a movement marks a shift away from industrialized, commodified building systems toward more place-based, ecological, and human-scaled building.

On the various styles of natural building, there are a wide variety of written materials detailing how to build using straw bales, adobe bricks, cob, rammed earth, earth bags, and pressed blocks, to name a few. This book is intended to fill a particular void: there is little literature published on *Light Straw Clay*. LSC is versatile, can be implemented in nearly any building project, and is compatible with nearly any superstructure. You could think of it as the "duct tape" of the natural building world. Much like flour and water can create a cornucopia of baked goods, straw and clay can be combined in myriad ratios and for a plethora of applications: floors, walls, ceilings, sculptures, new construction, renovation, interior walls, and exterior walls. It's insulative, fire resistant, durable, insect-resistant, it provides thermal mass; and the list goes on. The sky is the limit with LSC, and my hope is that this book inspires you to incorporate it into your project, building repertoire, and/or your own home.

Chapter 2

Rationale

What is light straw clay? It is an infill system of loose straw lightly coated in clay that is packed into forms (temporary or permanent) to insulate and enclose a building. It is not load bearing, but it can be mixed to a variety of densities. The "light" in light straw clay refers to two things: the amount of clay (less than in traditional daub in wattle and daub) and the material's weight (it is very light). Light straw clay is the English translation of the German word "leichtlehmbau."

Light straw clay was developed in Europe as an evolution of the wattle-and-daub infill system that filled in the spaces between structural members in half-timbered houses from the 12th century, on. According to Robert and Paula Laporte, North America's preeminent light straw clay experts:

> The light version was introduced after World War II, in Europe. The modern examples that Robert studied were infill systems like the wattle-and-daub predecessors, in which the light straw clay mixture was placed between a post-and-beam framework. In order to meet modern-day energy demands, this "lighter version" doubled the wall thickness and halved the density by introducing more straw, resulting in a radical increase in thermal performance.

There are a few other frequently used names for light straw clay: *light clay, straw clay* and *slip straw* are commonly used, and it has also been referred to as *rammed straw* (akin to rammed earth wall systems that have little-to-no fiber

Daub is a mud mixture applied to wattle *— a network of woven sticks in the wall.* Photo credit: Michael G. Smith

14th-century building uses wattle-and-daub infill. Wattle and daub is "heavy straw clay" — a predecessor to LSC. Photo credit: Robert Laporte

in them). For the sake of consistency, all of the above will be referred to in this book as *light straw clay,* or *LSC.*

Advantages of Light Straw Clay

As a light straw clay wall system is composed of more fiber (straw) than binder (clay), it is much lighter and much less dense than other natural wall systems like cob, adobe, or rammed earth.

Nontoxic

LSC is essentially nontoxic. Clay is known for its ability to absorb toxins due to its relatively large surface area and negative ionic charge. Because of this negative ionic charge, it absorbs positively charged ions which most bacteria, viruses, fungi, and toxic chemicals have an abundance of. (From "Living Clay," livingclayco.com) Combined with straw (which has minimal pesticides and herbicides applied to it as a crop, with only two applications being applied per year and not near harvest), it creates a nearly nontoxic wall insulation and enclosure. Being made mostly of cellulose, it is a relatively benign material and, like clay, it does not off-gas any volatile organic compounds (VOCs), unlike many other manufactured types of wall insulation.

Low-Tech Method and Materials

In addition to being a building material that is not hazardous to your health (although working with straw and dry clay can be very dusty, so it is best to wear a respirator), it is also very "low tech." The mixing and manufacture of light straw clay can be done without electricity: either manually or with a tumbler. Because the material is very lightweight, you don't have to be a muscled construction worker to participate in the installation. It can quite literally be fun for the whole family — including grandchildren and grandparents. It can be pleasant work; the job site can be quiet, the tools are very simple, and the material is all lightweight, nontoxic, and easy to install.

High Vapor Permeability

Light straw clay walls are highly vapor permeable. This is a benefit because vapor permeable walls allow moisture to exit a building instead of being trapped in wall cavities. Breathing, showering, cooking, etc., all generate water vapor in a building. Naturally "breathable" (aka vapor permeable) wall systems allow moisture, in the form of vapor, to move through the walls without the need for mechanical ventilation. Combined with clay-based binders and a clay-based plaster, an LSC wall system can help regulate humidity in a building, creating a much healthier environment in terms of air quality.

Vapor barriers are discouraged for LSC wall systems because they trap moisture moving through the natural wall system, where it can condense and be trapped. If water in a wall system can't escape, mold can form, which can be toxic for inhabitants. Clay's hygroscopic capacity allows it to absorb high volumes of water vapor and then release it — "breathing" it out — without compromising the straw or wood that it is in contact with. So, LSC in general, thanks to clay's properties, will regulate dampness, which also helps maintain its insulative qualities.

Place-based Architecture

The reason travelers find many types of vernacular architecture so appealing is partially due to their unique expression of people in a place. A building can be a reflection of place, and light straw clay buildings can embody place-based architecture by using materials from a structure's own site. One of the most common materials to use from a site (besides wood from harvested trees) is the earth excavated to make room for the foundation of a building. The amount of excavated earth may provide an adequate amount of clay for both wall infill and plaster skins. What better way for a building to fit into its environment than being built from materials found on the site?

Carbon Sequestration

Light straw clay being predominantly made of straw has the capacity to sequester carbon. Straw is made of cellulose; the chemical formula for cellulose is $C_6H_{10}O_5$, making straw approximately 45% carbon. If a building uses 100 2-string bales and each bale contains approximately 19 pounds (8.6 kg) of carbon, that is about 1 ton (860 kg) of stored, or sequestered, carbon.* This, in conjunction with earthen plasters, energy efficient design, and low impact living can substantially lower the carbon impacts of building a new residence. (**Calculations from Stuart Staniford, "Carbon Off-Set Values of Straw Bale Buildings," Aug 2010, earlywarn.blogspot.com*)

Chapter 3

Appropriate Use

OF THE MANY NATURAL WALL SYSTEMS to choose from, there are many reasons to choose a light straw clay wall system. Straw clay is highly compatible with framed wall systems because it is a non-load bearing material. Light straw clay can be infilled in nearly every wall framing system, be it timber framing, pole framing, conventional lumber framing, or framing specifically designed for straw clay infill.

LSC is also excellent retrofit insulation because preexisting walls can be furred out to any thickness. Furring out a wall simply involves adding stud material to the desired depth of wall (see "Larsen Truss System" in Chapter 6 for more on this). This can be done to the interior of a building or to the exterior. Using staggered studs or Larsen trusses also improves the insulation's performance because it allows the creation of a continuous thermal envelope, virtually eliminating the thermal bridging that occurs in a conventionally framed building (where solid studs create breaks between insulated stud cavities).

2×4 interior walls are infilled with LSC to aid in sound dampening and creating a sense of privacy for this small 800-square foot three-bedroom home. PHOTO CREDIT: LYDIA DOLEMAN

Interior walls can be infilled with straw clay in buildings that have exterior wall systems of other materials. Interior walls can benefit from the soundproofing and thermal mass that straw clay provides, and they provide a seamless look because they take plaster as well as other natural wall systems. If done with good and consistent formwork and with attention to detail, the walls can be very flat, lending themselves to very smooth finish plaster, which leads to less "dusting" through the life of the wall.

LSC's compatibility with conventional framing systems makes it easier to find contractors who can provide straightforward estimates for a project.

Wall systems or walls with lots of openings, like the south side of a passive solar building in the Northern Hemisphere, are highly compatible with straw clay, whereas cob, adobe, and straw bale are hard to work with around windows, doors, and other openings. It's a somewhat common practice to design a building that takes advantage of the high R-value of straw bales for the north, west, and east walls of a building, but use LSC in the south wall, which has the bulk of the glazing (windows).

One of the advantages light straw clay has over cob and adobe and other natural wall systems is that it slumps and sags very little while being installed, allowing an entire wall cavity to be filled in one work session. As long as tamping is consistent and there are not long periods of drying time between installations in the same wall/stud cavity, there is also very little shrinkage.

Straw bales are load bearing in this building that uses LSC for the south wall.
PHOTO CREDIT: LYDIA DOLEMAN

Even with the intense heat application of a blow torch, LSC only smolders.
PHOTO CREDIT: LYDIA DOLEMAN

Many projects, particularly in urban areas, have to be carried out in limited space. When there is limited square footage to work with, the 18″ to 24″ width of straw bales or cob may rule these wall systems out because they eat into usable space. In urban areas and sites with limited space, straw clay can be an excellent choice to create thinner wall systems that are still highly insulative. LSC can be made to fill any wall width that can reasonably dry within the timeframe of most building seasons. Most LSC walls do not exceed a 12″ thickness.

Straw clay is very fire resistant. Tests conducted by Joshua Thorton and John Straube found that, based on ASTM standards E 111 and E 84, LSC would very likely meet the conditions required for a fire-resistant period of four hours. They also reported that LSC is a "highly ductile material with the potential to absorb a fair amount of energy in the event of seismic activity." (Thornton, *Initial Material Characterization of Straw Light Clay,* Canada Mortgage and Housing Corporation, 2005.)

As each piece of straw has been coated in clay and packed in the wall, there is very little that can actually combust. Although, the walls are breathable to vapor, the continuous wall envelope should not have open channels for sufficient quantities of oxygen to be present, which also helps LSC resist combustion. Like a lot of dense materials, it may only smolder.

According to architect Franz Volhard, one of the European leaders of earthen construction methods, his own fire tests of LSC demonstrated:

- Light earth responds passively to the effects of flames, i.e. it does not contribute to the spread of fire.
- The formation of an "insulating" charred layer protects the surface of

underlaying materials from direct exposure to the flame, which increases with flame duration.

- Neither smoke, nor fumes nor perceptible combustion gases were produced.
- No particles fell from the specimens which could have contributed to the spread of the fire.
- Compared with wood-wool magnesite-bonded panel, the fire behaviour was better with less charring and no smoke development.

These results suggest that straw light earth could be classed as B1 "Not easily flammable."

(Franz Volhard, *Light Earth Building: A Handbook for Building with Wood and Earth,* 2016, p. 225.)

Given that the required formwork for LSC is usually very flat and rigid and resists bending under pressure (plywood with "strongback" reinforcement), it is easy to form flat walls into very rectilinear structures, be it craftsman-style bungalows or modernist buildings. Light straw clay can also be made into bricks. Once dry, they can be placed directly in wall cavities, stacked to insulate thick walls in wetter climates (where there wouldn't be enough time in a building season for LSC infill to dry), or cut and fit in to odd spaces that need insulation. Curved walls and window openings can also be made with light straw clay, although they require more detailed planning and curved formwork.

Due to the nonstructural aspect of light straw clay, a superstructure must be erected prior to installing the wall insulation. Therefore, you need a dry work space, in most situations. There are various methods of infill, some of which leave the roof framing until after the walls are filled in. In wet climates, having a roof structure

Rounded corners are achievable with LSC wall systems by using rounded forms.
Photo credit: Lydia Doleman

Rounded corners in the making.
Photo credit: Jim Rieland

up allows more time for installing the LSC and is added insurance should a storm happen during an anticipated dry season. However, having a roofless building allows for significantly more airflow to help facilitate drying.

Other applications of light straw clay are ceilings and floors. LSC can be used between floor joists if permanent formwork is used, or by implementing what is referred to as "light earth reels" (Volhard) or "chorizo style" (they look like sausage rolls), where the straw clay is wrapped around a stick or bamboo rod and suspended between the floor joists. Adequate airflow must be provided, as it requires a much heavier clay slip to wind the straw and clay around a stick that will adhere while being suspended overhead. In addition to all the other advantages of LSC, the advantage of this method is that it takes plaster very well; plus, it can be very artistic.

A board-and-batten rainscreen keeps the wind-driven rain from saturating the LSC on the south side of this handsome brewery. Photo credit: Dean Hawn, Burning Daylight Construction

A half rainscreen keeps the bottom portion of this tiny timber frame LSC from being damaged by splashback from the roof. Photo credit: Lydia Doleman

If you live in an area with wind-driven rains, you might want to consider a rainscreen to protect your finished walls. Wind-driven rain does not rule out building with LSC, but it is best to plan for worst-case scenarios. This is also true for areas of high snow. A snow drift next to an earthen plaster or lime-plastered wall will cause problems. A rainscreen (even for a half wall) can be a smart choice to mitigate climatic issues.

Other Important Considerations

If you are considering building a full-sized residence or something over 800 square feet, consider building a small shed or outbuilding first. It's much more cost effective to work the kinks out on a small building than on a big one. There are so many little details in an LSC building. Things will be much more streamlined if you are familiar with its idiosyncrasies, so it's best to start with a smaller project rather than jumping right into a big one. You may even discover that you don't want to use LSC on the larger structure, or that you may want to use it in only particular parts of your building. Starting small will be much less stressful, too; you won't feel the same time pressure about the myriad daily decisions you have to make on a construction project if you have gone through them once on a smaller project. Plus, you will have the added bonus of having the smaller structure as a place to put all your tools (or yourself!) while you work on the bigger one.

(Due to code constraints and permit costs, a lot of people choose to build smaller structures under inspectors' radar. But be careful with this.

Small "outbuildings" are often characterized by code as sheds, or non-inhabited agricultural buildings; often they are not required to meet code, but different municipalities have different requirements, so it is important to thoroughly look into your local building codes.)

Disadvantages of Light Straw Clay Wall Systems

Labor

Light straw clay is labor intensive. Even with a tumbler (more on that later), you will need a group of people to pack the straw into the wall cavities and move forms. It takes a minimum of five people to mix slip, mix straw clay, and pack it into walls. You could do this all solo; but it would take you several seasons for a whole house. But try turning this disadvantage into an advantage. Take the more-the-merrier approach, and turn your worksite into a work party! Remember that the labor can be done by practically anybody, and it's relatively quiet and pleasant. (See Chapter 8 for more on work parties).

Dry Times

The rule of thumb with straw clay is that you need approximately one week per inch for dry time. For the average 12-inch-thick wall, that's 12 weeks (more incentive to get the walls done quickly). During that time there are a lot of other parts of a project to complete, but especially in locales that have short building seasons and extreme weather, you must plan ahead for adequate drying time. With longer dry times, mold can be an issue. Little white fuzzy spots of mold sometimes appear, especially in areas with low air circulation or where an LSC mix with a higher proportion of clay was used, e.g., tops of stud cavities and repairs. (This light white mold is harmless.) If using a water thinner, like borax, in your slip, it can cause efflorescence, which is a buildup of salts that crystallize on the surface as they dry. These are harmless, and can be gently scraped off.

The amount of moisture dissipating from a straw clay wall system can sometimes overload an interior and cause surface molds to appear on wood, especially if the wood wasn't completely dry, or it wasn't treated/sealed. The use of fans and dehumidifiers is highly encouraged. Propane heaters are *not* recommended; they produce a fair amount of moisture in their heat production.

In systems where the formwork is permanent, there is less exposed surface area for the LSC. In situations like this, or when dry time is short and/or humidity is high, care must be taken not to make too heavy a clay mix; you want to allow the LSC to dry well in the closed wall cavity. The use of water thinners to limit the amount of moisture in the slip is advised. For more information on what, when, and why to use water thinners, see Chapter 5, "Water." Once again, the use of fans and dehumidifiers is key to facilitating the drying of an LSC wall system.

Specific Exclusions

There are some situations that are *not* right for straw clay. You should consider other options if you:

- live in an area with an excessively short building season (one where wet weather is followed closely by freezing weather)
- have no access to materials for building the supporting structure
- have limited access to straw or clay
- have no labor pool
- live in a building with preexisting termite or mold issues
- have a major straw allergy.

These are some of the reasons you might want to explore a different wall system. Note, however, that some innovative builders have implemented the use of straw clay *bricks* for short-season building. The increase in surface area greatly decreases dry times, and the bricks are light and stackable.

Some regions in North America do not have a code for the use of straw clay as an insulating wall system (See Chapter 11 – Building Codes). This could make a permitted project challenging because you would have to work with the local building department for permission to build this way. Also, some banks may not grant a loan for an unconventional wall system; and some insurance companies will not insure an unconventional project.

Properties

The most obvious property of LSC is its insulative ability. Insulation levels are rated by their "R-value," which is the measure of a material's resistance to heat flow. A higher R-value means better thermal performance. The R-value of LSC walls is mostly contingent on the density to which it is packed and the amount of clay used. Good tests have been done on the R-value of LSC by the Design Coalition, Douglas Piltingsrud, Franz Volhard, and others.

Note that in the testing done to produce Tables 3.1 and 3.2, the amount of straw in each mix remained consistent; only a change in the amount of subsoil in the mix impacted density and, therefore, the thermal resistance (R-value) of the LSC. In short, less subsoil equals a lighter LSC mix and a higher R-value per inch, while more subsoil equals a heavier LSC mix and a lower R-value per inch.

The R-value of a wall assembly will also depend to some extent on what type of wall system is in place. If building a structure and not using split stud or Larsen trusses, the thermal envelope will not be continuous; the insulating quality of the entire wall system is diminished because there will be thermal bridging through the studs.

Optimal Density

Given the broad range of densities achievable with LSC, it is important to achieve the correct one to meet the desired R-value for a wall system. Field experiments are required to find the ratios that will yield your desired density.

The following is a guideline recipe by Douglas Piltingsrud (designcoalition.org):

Following is a basic formula for making a wall density of 13 lbs/ft^3 (pcf) with a thermal

Table 3.1: Thermal resistance of straw-clay at different densities

Specimen	Density (#/ft^3)	Density (kg/m^3)	Conductivity (W/m°K)	Delta Temp. (deg. C)	Temp-median (deg. C)	R/inch (hr°F·ft^2/BTU/inch)
Low Dens. I	10.2	164	0.08	4.54	23.21	1.80
Low Dens. II	13.0	209	0.09	approx 4.5	approx 23	1.69
So. Dakota I	15.8	254	0.09	4.51	23.40	1.55
So Dakota II	13.3	213	0.09	4.54	23.39	1.67
Reg I	13.3	213	0.08	4.55	21.30	1.72
Reg II	13.7	220	0.09	4.54	23.59	1.66
NM I	38.1	612	0.13	4.58	23.21	1.11
NM II	43.9	705	0.16	4.19	23.40	0.90

Credit: Design Coalition

Table 3.2: Properties of light straw clay mixtures (1)

Lighter → Heavier

Density pcf (kg/m³)	Straw pcf (kg/m³)	Subsoil pcf (kg/m³)	Water gal/cf (l/m³) (2)	Min. % Clay in Subsoil	Min. Clay Salt Ratio	Subsoil Testing Method (3) (4)	Subsoil Volume % (5)	Max. Wall Thickness Inches (mm)	Straw Volume % (6)	Air Volume %	Straw + Air Volume %	R-value (hr•°F•ft³ • BTU per Inch Thick)	Conductivity (W/m°K)
10 (160)	6.7 (107)	3.3 (53)	1.55 (208)	70	3.5 : 1	A	2.0	15 (381)	7.4	90.6	98.0	1.80	0.079
12 (192)	6.7 (107)	5.3 (85)	1.63 (218)	46	1.7 : 1	A	3.1	15 (381)	7.4	89.4	96.9	1.72	0.084
13 (208)	6.7 (107)	6.3 (101)	1.67 (224)	40	1.33 : 1	A	3.7	15 (381)	7.4	88.9	96.3	1.69	0.086
15 (240)	6.7 (107)	8.3 (133)	1.74 (233)	35	0.95 : 1	A	4.9	15 (381)	7.4	87.7	95.1	1.63	0.090
20 (320)	6.7 (107)	13.3 (213)	1.93 (258)	30	0.60 : 1	A	7.9	12 (305)	7.4	84.7	92.1	1.48	0.101
30 (481)	6.7 (107)	23.3 (373)	2.31 (285)	N/A	N/A	B	13.8	12 (305)	7.4	78.8	86.2	1.22	0.122
40 (641)	6.7 (107)	33.3 (533)	2.70 (362)	N/A	N/A	B	19.9	12 (305)	7.4	72.8	80.2	1.01	0.143
50 (801)	6.7 (107)	43.3 (694)	3.08 (412)	N/A	N/A	B	25.7	12 (305)	7.4	66.9	74.3	0.84	0.163

(1) Values in the table may be interpolated

(2) Water mixed with subsoil equals clay slip

(3) Subsoil Testing Methods per 2015 IRC Appendix R Commentary (also upcoming in 2018 version...)

A. Lab test for percent of clay, sand and silt via hydrometer method

B. Ribbon test or the Figure 3 Ball Test in the Appendix of ASTM E2392/E2392M

(4) Trace amounts of organic materials are acceptable

(5) Uses 168 pcf for subsoil specific gravity (2.7 g/cc)

(6) Uses 90 pcf for straw specific gravity (1.45 g/cc). Straw volume % and associated columns may increase for 20 pcf walls and above due to the weight from additional subsoil.

CREDIT: DOUGLAS PITINGSRUD AND STRAWCLAY.ORG

resistance of R-1.69 per inch. A 12-inch (30 cm) thick wall with this density will yield an insulation value of R-20. At a 15-inch (38 cm) wall thickness, the R-value would be 25.4.

More information on R-value is provided in Chapter 4, Building Science Notes.

Prior to commencing the LSC wall infill, it is important to figure out what quantities of clay are needed to achieve the wall density you are looking for.

Example: 13 pcf/R-1.69 per inch Wall Density Formula

- 30 lbs of straw (13.6 kg)
- 28 lbs of subsoil (12.7 kg) with minimum 40% clay content
- 58.4 lbs of water (7 gallons or 26.5 liters)

Creating Test Blocks to Assess LSC Mix Density

Build a minimum of three boxes that have one cubic foot in their interior. Use screws to put it together on at least one side, so you can remove a side of the box to assess compaction and remove the sample.

Using gallon buckets or a similar scalable vessel, and starting with the slip formula above for an optimal mix of 13 lbs/ft³, make some test batches. If you are using site subsoil with an unknown percentage of clay instead of bagged clay, then make a number of batches using different amounts of subsoil and ranging from a heavy slip to very light slip, then mix each of the slips with 6.7 pounds of straw, and tamp the LSC into the forms. Three forms can give a very good range.

Label your forms, so you'll remember which mix has which amount of subsoil. Remove one side of the form after tamping to make sure compaction is suitable. Remove the straw clay sample and let dry. Expedite this process with heat and air movement (sun is great too, but protect from rain) to dry the samples. Once dry, you can weigh the samples to find the achieved densities.

You can then evaluate both the variability of the density of the mix based on the amount of subsoil in each sample and the variability of the density of the sample related to the level of tamping.

An optimal mix will yield a dry sample of 13 lbs/ft^3, which is R-1.69 per inch. If you have a lighter sample, you will have a higher R-value, but you may be losing the necessary binder by having too little subsoil in your slip or insufficient tamping. If you have a heavier sample, then you will have a lower R-value, which means that the subsoil quantity is higher than optimal and/or you are compacting the mixture too much. With testing and practice, it is possible to achieve an optimal density of mix and to get a feel for the correct level of tamping. Other desired densities and R-values can be achieved using the same process and varying the amount of slip and compaction.

Soundproofing

No testing has been done yet on the acoustic qualities of an LSC wall system. However, Franz Volhard calculates sound reduction by "using the values of other massive building elements with a corresponding mass per unit area." He adds: "Earth building materials add mass to a timber frame structure and it is possible to achieve good sound insulation using a simple, single-skin construction. Compared with other massive wall infill materials — all earth materials and light earth in particular are softer and more elastic. Sound vibrations are softened and attenuated. As a heavy but soft building material, earth therefore offers excellent sound insulating properties." (Franz Volhard, *Light Earth Building,* 2016, pg. 227–228)

Given the variability in density, the potential for split-stud construction and the variety of skins that are compatible with light straw clay, the potential for sound absorption is high.

Chapter 4

Building Science Notes

MOISTURE, TEMPERATURE, WATER VAPOR, and air movement are all factors to be considered and designed for in any building that is built to last and perform well. Given the organic nature of light straw clay wall systems, the need for breathability, and the general desire most people have to inhabit a comfortable space, it is wise to apply the principles of building science to your building to ensure a long-lasting, relatively low maintenance, and well-insulated structure.

Use the basic principles of building science to manage four key factors in a building:

- Moisture
- Temperature
- Water vapor
- Air movement

Principles for Thermal Control

Much like the osmosis of salt water to fresh water, the movement of heat is always from hot to cold. Touch a cold rock, and that is your body heat transferring to the rock; what we call "cold" is really just the absence of heat. There are three different ways that heat moves: radiation, conduction, and convection.

Radiation

Just as it sounds, radiation is heat energy that emanates from something; it can do this as a wave or as a particle. The sun, heat lamps, and our bodies are examples of things that radiate heat. If an object is cooler, it will absorb that energy. If the rock just mentioned were set in the sun, it would absorb solar radiation and heat up. A passing cloud could block that radiation — just as blinds and roof overhangs keep the sun's radiation (aka heat) out of buildings.

Conduction

Conduction is when heat energy is transferred through direct contact. Take a thermal image after you touch that cold rock, and you will see where your hand left some of its heat energy. The more conductive a material is, the more heat it will transfer. Generally, conductivity is a function of density. The less dense a material is, the less conductive it is. Put a pair of gloves on and the thermal image will reveal little heat transferred to the cold rock, because the glove is much less dense than your hand.

Convection

Convection is the movement of heat through fluids (doesn't happen in solids) where the warm fluid (generally, air, in the context of buildings) changes density as it gets warmer and more diffuse, and rises, and then dissipates its heat energy, becoming cooler and denser, and therefore falling. Convection is important to control in buildings by controlling the amount of air that is free to circulate convectively. If your LSC wall assembly has big gaps or voids, convection can occur — having a negative impact on the efficiency of the insulation.

Insulation in a building is the thermal control layer. It is the barrier to the movement of heat from interior to exterior and vice versa. The amount a wall assembly resists heat flow is measured as *R-value* (resistance to heat flow); the other way it is commonly measured is by the static-state conductivity, or U-value. Building codes specify the values particular building materials must have, insulation and windows specifically, and they are different for wall assemblies, attics, ceilings, and foundations.

Most wall assemblies are an assemblage of different materials — all with different rates of conductivity. In standard construction, where the wall is broken up into stud bays, the studs have significantly lower resistance to heat flow than the fiberglass, cotton, or whatever insulation materials are used. This causes "thermal bridging"; you can see it demonstrated on a cold morning in some rooftops because the snow melts along the roof joist lines, but not above the insulated sections of the space. Less insulative materials, settling of fluffy insulation, and improper keying that results in drafts between materials are other ways heat is allowed to escape, lessening the insulative performance of the whole building.

R-Values and Light Straw Clay

Research conducted by the Design Coalition in 2003 revealed a range of R-values for LSC, related to the density of the material. The range was an R-value from 0.90 per inch to 1.80 per inch with a factor of nearly 4 in the difference in densities.

A low-density mix in a 12″ thick LSC wall provides more than the R-21 commonly required by codes for wall insulation. Given that LSC wall systems generally incorporate a continuous thermal envelope (negating the common "thermal bridging" present in conventional construction), they perform better than a standard building with R-21 batt insulation. "Thermal bridging" is an area in the building envelope that has significantly higher conductivity than adjacent, more insulating materials. As mentioned above, in an average conventionally framed house, the studs conduct heat at a faster rate than the fiberglass insulation in the stud cavities, therefore reducing the overall insulative qualities of the wall system. The 2015 International Residential Code (IRC) recognizes LSC as having an R-value of 1.6 per inch without relation to density. (See IRC section AR104.1 *R-Value.*) Revisions approved for the 2018 IRC tie R-value to density. LSC buildings also generally have a plaster skin on one side (or both), increasing the capacity for a wall system to store heat and radiate it out later, because plaster is generally more dense than standard ½″ sheet rock and is made of denser materials, like sand. The straw clay itself is also much denser than common insulating materials.

1. Studs
2. Insulation
3. Siding

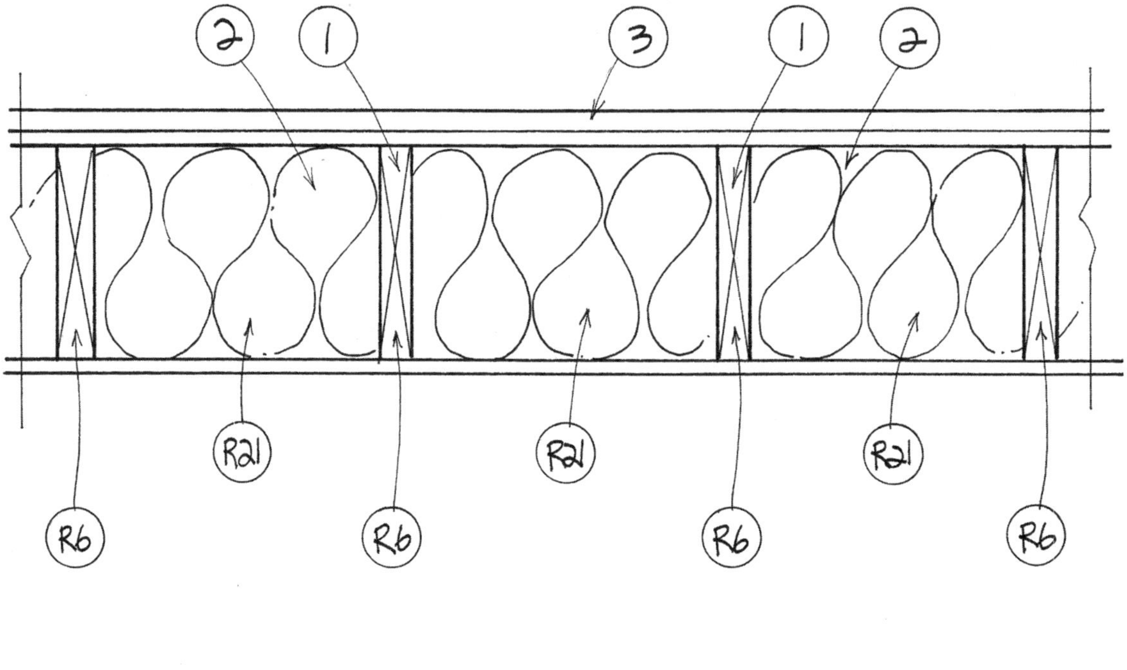

Air Control

The control of air moving through a wall assembly is essential for a building to perform well. A small draft is easily felt, and air movement in and out of a building can dramatically alter the thermal performance of an otherwise well-insulated building. Air is controlled with a variety of layers that are designed to prevent air leakage. Just as you don't want your roof to leak rainwater, you don't want your wall assembly to leak air (or rainwater), either. Detailing around wall penetrations is very important. Where plumbing, gas, and electrical enter a building from the exterior or unheated spaces into heated spaces, it is important to adequately block air from leaking through these openings; in addition, you need to seal the gaps between building elements: floors to walls, and walls to ceilings and roofs.

Air Control in LSC Buildings

Due to the need to maintain adequate vapor permeability for natural wall systems, vapor barriers are generally frowned upon for LSC wall systems. The IRC appendix for LSC requires a moisture barrier only between the bottom of the walls and any masonry or concrete. It requires a coat of plaster as an air barrier even when exterior siding is used, as well as a ventilating airspace between the plaster and the siding for adequate moisture diffusion.

The primary air barrier in an LSC wall system is the plaster skin. It should be continuous, crack free, and is applied in a way that prevents air leaks. Channeled trim with a bead of caulk, J-channel over air fins, and plaster as a cover over dissimilar materials are some of the things that keep the LSC wall envelope continuous and from leaking air (and rainwater).

Although there is only a minor amount of shrinkage that occurs when the clay in the LSC wall cavities dries, there is *some* shrinkage. Thus, it is possible to get hairline cracks along the top and edges of walls (especially in conventionally stud framed, noncontinuous thermal-envelope-style construction). This is where appropriate keying and air fins are vital to control the air flow in the wall system. (See Chapter 7, "A Note on Air Fins," for more details.)

The transitions between ceiling and attic and wall insulation are particularly vulnerable spots for air leakage and need special attention. Partly this is because you have so many dissimilar materials coming together. It is also an area in the building where warm, rising air is apt to push out of a building (which means that in the cracks at lower elevations, cold air will be pulled in due to the natural convection cycle that occurs in a heated space).

Air and Moisture Control = Vapor Control

The hotter the air, the more moisture it can hold (water as vapor). As air tries to attain a temperature equilibrium (seeking cold), it will cool, and its ability to hold the amount of moisture it once had when warmer will decrease, and the vapor will condense into water droplets. An LSC wall system can handle a high volume of vapor due to clay's ability to absorb lots of moisture, but once vapor has condensed to water it can cause serious issues in a wall system of any type. Even a small crack can conduct a surprising volume of moisture through a wall. Therefore, controlling air flow is also a component of vapor control.

Moisture enters a wall through the process of diffusion: water vapor molecules move from areas of a high concentration through pore spaces in a material to areas of lower concentration. This process is measured in "perms" i.e., the ability of a material to resist the diffusion of moisture.

There are three ratings/classes for a material's ability to impede the passage of vapor or act as a vapor retarder:

- Class I — 0.1 perm or less (qualifies as a vapor barrier, or vapor impermeable)
- Class II — 0.1 to 1.0 perms (vapor semi-impermeable)
- Class III — 1.0 to 10 perms (vapor semi-permeable)

(From Chris Magwood's *Essential Hempcrete Construction,* p. 19.)

Due to the high clay content in LSC wall systems, vapor can readily move through the wall assembly and should not be constrained by a vapor barrier in direct contact with the LSC. But the system will fail if there are voids in the wall assembly. These can allow condensation to form somewhere in the middle of an LSC wall.

Water Control

Point loads or "bulk loading" of moisture is one of the biggest threats to any natural wall system. Light straw clay can lose its insulating capacity if it gets wet. Mold can also be a problem, but generally won't result in a structural issue if only the LSC is receiving the water. If there is a chronic water issue — leaky plumbing, gutter leaks, poor window detailing — this can result in structural problems and more serious mold issues.

The design principle of creating a "good hat, a good coat, and good boots" for a building is part of the strategy to protect walls from water infiltration. Generous overhangs and a minimum of 8″ clearance for the LSC and its plaster above grade (an IRC requirement) are also part of keeping a building dry. Rainscreens are an excellent way of protecting the entire exterior of a building from wind-driven rain. Sometimes only the sides that are predominantly exposed to the weather need rainscreens.

Generally, water control is about controlling exterior-driven moisture — keeping rain and snow away from a building. Plasters need to be designed to be durable and compatible with climactic conditions. In areas with high wind-driven rain, earthen and even lime plasters may not be able to handle the amount of water and its erosive effects. Water that gets into cracks in plasters and then freezes can pop off whole sections of plaster.

Rainscreens are an attractive and functional way to handle rain, as they keep the water away from the LSC and allow airspace for wind-driven rain to drip out and vapor to exit the wall assembly.

Good, preventative design can help mitigate damage in the event of an interior water overload from broken plumbing, excessive showering, condensation on windows sills, etc. Holding plaster skins and base trim slightly above floors can help prevent the wicking of water up into the wall. A double bottom plate can also help prevent water from actually reaching the LSC insulation. Chronic condensation issues that stem from overloading a system with vapor may require mechanical ventilation. Bathrooms and kitchens usually have some required form of ventilation. If you are constructing a building with the intention of a creating a very tight building envelope, mechanical ventilation will be necessary.

Chapter 5
Material Specifications

Straw

STRAW CAN BE FOUND on six of the seven continents and is common in most rural areas. It is even available in most urban areas (look for feed stores). In rural areas, if asking your neighbors doesn't lead you to someone with straw, check with the local feed stores or garden centers. Ask if you can talk with their suppliers; you can save a lot of money if you buy directly from the source — the amount can be substantial. Depending on the thickness of walls, even a small 120 square foot structure can require more than 15 two-string bales. A full-sized residence might require more than 200.

Types of Straw and Straw Bales

Straw comes in a variety of flavors and shapes. When buying straw for a project and calculating how much you might need, it is really important to note the difference between 2-string, 3-string, and 5-string bales, and round bales. Two-string bales are straw bales that are baled using two polypropylene strings (rarely, wire these days) that are usually 14–18″ deep, by 14–18″ high, by 36–48″ long; they weigh 30–60 pounds, depending on how tightly they were baled. Sizes vary, though; make sure to ask your dealer about the size of the bales on offer. It is crucial to get the actual dimensions prior to buying your bales for a project, as your estimate will be based on volume per bale.

Three-string bales are generally 24″ deep by 18″ high by, 33–56″ long and can weigh up to 100 pounds each. Five-string bales require a tractor to move and contain about a ton of straw. Generally, most people get the 2-string bales for ease of moving — you will likely move them twice before breaking them open to use in the light straw clay mix. If you live in an agricultural area, 3-string bales tend to be the most available, as they are the most efficient way to get straw out of the field.

Stack bales in an interlocking manner so they are stable.
PHOTO CREDIT: LYDIA DOLEMAN

A well-baled bale will only have about 2–3 inches of give in the strings. Be sure to check; loosely baled bales not only mean less straw per bale, they can also fall apart during transport. If you get a chance to visit the farm or to visually inspect the bales, it is good to see a broken-up one just to gauge what length of straw is in

Table 5.1: Environmental impacts of straws

Ecosystem Impacts	Embodied Energy	Carbon Footprint	Indoor Environment	Waste
Low to Moderate. Impacts largely the result of monoculture agriculture, including fertilizer, herbicide, and pesticide use. Confirm practices with straw bale source to verify degree of impacts. Straw from organically grown crops will have the lowest impacts.	**Very low.** 0.24 MJ/kg* or 3.5–4.0 MJ per average two-string bale. No high heat processes required. Production energy input split with embodied energy of cereal grain production.	**Very low.** 0.06 kgCO2e/kg* or 0.2 kgCO2e/kg per average two-string bale. Production carbon output split with carbon output of cereal grain production. High carbon sequestration potential.	**Very low to Low.** Very low surface toxicity. No toxic off gassing. Material separated from interior air by plaster or sheathing.	**Very low to Low.** Construction: Leftover or unused straw is fully compostable. Polypropylene strings may be recycled in some jurisdictions. End of life: Straw is fully compostable. Embedded mesh will require separation.

Note:
* Data is from Inventory of Carbon and Energy (ICE) 2.0, University of Bath.

A tightly baled bale will have little give in the strings, which indicates plenty of straw in the bale. Loosely baled bales contain less straw and tend to break open while being moved. This is a well-baled bale! Photo credit: Lydia Doleman

the bales. Some bales are composed of short shredded straw, with straw lengths only about 4 inches long. This is still usable for straw clay, but it's less than ideal, because it will soak up more clay than longer straw would, adding to weight, clay use, and dry time.

Although composition will vary depending on what is grown in your area, your bales will most likely be one of the following: wheat, barley, oat, or rice. Rye, grass from grass seed production, and teff are also sometimes found. But there is currently no documentation of these types of straw being used in an LSC wall system. In the event that these are the only types available, purchase one bale and do a test.

Any fiber can be used in place of the "straw" in light straw clay (wood, hemp, animal fur, or paper, for example) as can minerals like perlite, vermiculite, pumice, etc. However, it is the hollow tubes in straw that provide the insulation, so without those, the R-value of your wall won't be as high. If you are in an area that gives you access to all four of the most common types of straw, it's best to know why you might choose one over the other. They each have pros and cons when being used for LSC.

Wheat, barley, and oats all have long very, tubular straw, which provides excellent insulation when coated in a clay slip — lots of air gets trapped in those tubes. The minor downside is that there is an oily substance in these straws that makes them require a little more clay slip than rice straw to properly coat them.

I have observed a little more seed head in these three types, which is only a minor issue. It is best to have very little to none to deter rodents and insects, although when all the seed heads sprout in the walls, it looks pretty cool (see an example in the color section), is a great conversation starter, and helps pull the moisture out of the walls. The wilting of the sprouts is an excellent indicator that the moisture is leaving the wall system, and it is nearly dry.

Rice straw does not have that perfect tube like the others; however, it has a higher silica content, which allows it to resist rotting in higher-moisture situations. Rice straw bales also tend to have longer fibers than wheat, barley, or oat bales. It might be a superior straw for late-in-the-season projects, or if you are looking to do chorizo-style straw clay. Rice straw bales are particularly dusty, so beware, and wear protective gear!

Straw is baled at a particular moisture content (always under 20% and usually between 8–13%). Bale moisture content should be tested with a bale moisture probe, like a moisture meter used for baled hay.

Water

It seems obvious, but an incredible amount of water goes into mixing up the slip and cleaning everything and everybody off at the end of the day. A rough guideline is that you will need 0.80 gallons (3 liters) of water for mixing the LSC for each cubic foot of wall. So, for a 1,200 cubic foot structure, 960 gallons (3,634 liters) would be needed for mixing; and you should triple that amount of water to have enough extra for leaks, spills, extra challenging cleanup, and for cleaning people.

This barley straw is resisting the slip.
PHOTO CREDIT: LYDIA DOLEMAN

For wetter or humid climates that have short dry times, add 1 cup borax to 40 gallons of slip. The borax "thins" the water, so you need less water to get the slip to the right consistency. This way, less water has to evaporate from your walls before they are dry. Borax also inhibits mold, insects, and fire.

Clay

Clay can be found in most places, both urban and rural. Most soils are varying proportions of sand, silt, clay, and organic matter. If you are excavating for your project, have your excavator clear the topsoil off and place it in a separate

Table 5.2: Environmental impacts of clay plaster

Ecosystem Impacts	Embodied Energy	Carbon Footprint	Indoor Environment	Waste
Very low to Low. Site soils can be excavated and used with minimal disturbance to site. Processed or bagged clay impacts will vary with extraction and refinement processes. Confirm practices with source to verify degree of impacts. Large-scale quarrying can cause habitat destruction and surface and ground water interference and contamination.	**Very low.** 0.083 MJ/kg* or 0.58 MJ/m^2 (for excavating and transporting soil locally). No figures available for bagged clay products, but figures will be higher due to energy used for crushing and drying clay.	**Very low.** 0.0052 kgCO2e/kg* or 0.108 kgCO2e/m^2 (for excavating and transporting soil locally). No figures available for bagged clay products, but figures will be higher due to energy used for crushing and drying clay.	**Low to Moderate.** Clay plasters typically contribute to high indoor air quality. Contaminants in soil or bagged clay may be difficult to assess and pose some difficulty in confirming air quality impact. Hygroscopic qualities of clay provide excellent moisture handling characteristics.	**Very low.** Construction: Leftover plaster can be returned to the ground with no impact. End of life: No ill impacts. Embedded mesh will require separation.

Note:
* Data is from Inventory of Carbon and Energy (ICE) 2.0, University of Bath.

pile; you can use that topsoil (which is mostly organic matter) in the compost pile, and later in the garden. The next layer to be moved is the subsoil. This is where you will find the clay. An excavation for a foundation will likely create a big pile of potential clay for you. That would be the first and most ideal place to get it from.

How do you know if your site soil contains clay? Clay is fantastically sticky and smooth. However, so is silt when it is wet. Silt is microscopic sand; when it is wet, the capillary action of water gives it a similar texture to moistened clay. But silt does not have the same adhesive properties as clay when dry. The easiest test to determine if what you have is, in fact, clay and not silt is the "worm" test. Get a handful of subsoil wet and mix it to a thick, putty-like consistency. With clay, the more you mix it, the more you work the water molecules between the microscopic platelets of clay, making it more and more plastic. So work that first handful well — for at least two minutes. Then roll it into a "worm" — a long tube about ½" thick. If you can get a worm, it is likely good clay soil. If you can take that worm and bend it 180 degrees, and it stays together, you have a very high percentage of clay. However, although your subsoil may pass this first test with flying colors, there is one more test.

Make a ball out of that worm and let it dry rock hard. If, as it dries, it cracks profusely and crumbles, you may have *expansive* clay, which, sadly, is no good for straw clay. It's good for lining ponds, but not for building because its molecular bonds break when it dries and shrinks. If, though, your ball has no major cracks, then take that ball and drop it from hip height. If it mostly stays together, then you can celebrate.

If you just have sand, too much silt, or expansive clay, you will need to source off-site clay. Good places to check are local building sites or Craigslist. Make a visit, do a test, and be sure to ask why they are getting rid of the subsoil.

The Worm Test

1

2

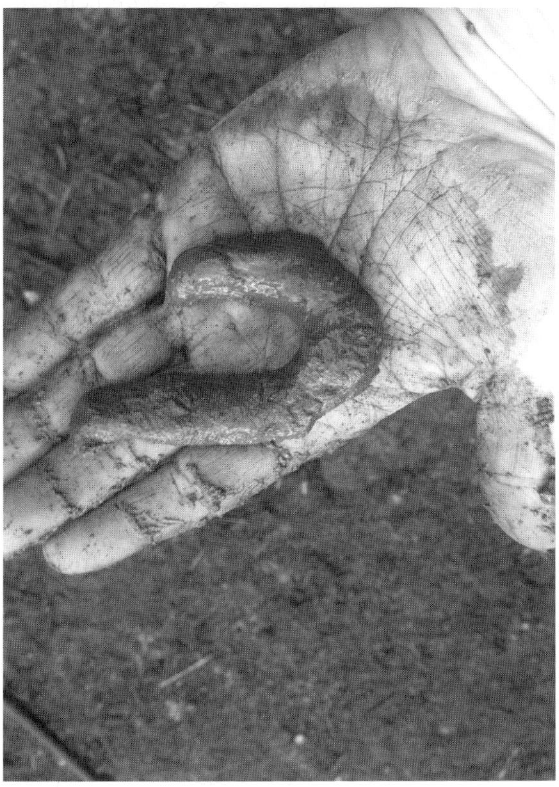

3a

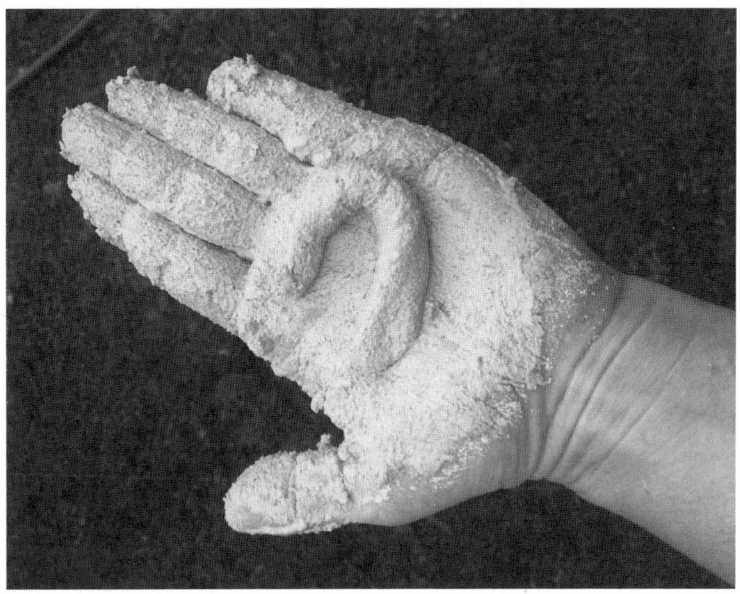

3b

1. Make a worm out of well worked clay dirt.

2. Bend the worm 90–180 degrees. The breaking shown here indicates too much sand and silt.

3a-b. Non-cracking 180-degree bend = good clay!

PHOTO CREDIT: LYDIA DOLEMAN

If it looks like an industrial site or a former gas station, it's likely toxic, and shouldn't be used for building. As a side note, when they can't use on-site subsoil, most of the builders in Portland, Oregon, have an excellent and very regular source of consistently clay-laden subsoil: the graveyard.

Sometimes just adding a little pure clay to lightly clayey subsoil helps immensely with its stickiness. For smaller projects, or a project where there is just barely enough clay present in the subsoil, clay can be purchased in powdered form from a mason's supply company or a ceramics supplier. *EPK*, or kaolin, clay is a porcelain clay, making it a wonderful light color for plasters, but it is very soft and dusty when dry. It can usually be purchased in 50-pound bags for anywhere from $7–40 per bag. One bag will make two 5-gallon buckets of slip, which is very little in the grand scheme of things — so bagged clay might not be your best option. On the other hand, if you consider the time and labor involved in sifting soil, bagged clay could be the way to go. If you are really on a budget and in an urban area, you could possibly source broken bags of clay from ceramics supply warehouses.

Stronger clays are known as *ball clay* or *fire clays*; if you have the option, go for these.

Another option (only recommended for a small project where no site clay is available) is to procure buckets of clay remnants from pottery studios and soak them. But it is remarkably time and labor intensive to get hard or putty-consistency clay back into a liquid state.

If you are determined to get your clay from on site, and you have a large property, look for areas where water pools after a rain. Water that doesn't drain through the soil could be an indicator that there is a lens of clay below. Take a shovel and investigate. If you strike clay, you will likely need to dig a pretty sizable hole (read: need heavy equipment) to acquire enough clay for an average-sized house.

Plasters

Plaster has been the traditional method of protecting LSC from the elements and providing an air barrier. It is the most common way of covering/protecting an LSC wall system.

There are two recommended types of plaster: earthen and lime.

Earthen Plasters

Earthen plasters are composites of clay as a binder, sand as structure, and fiber for tensile strength. Finish plasters often incorporate a pigment of some type. Other additives are covered in this chapter.

Earthen plasters are generally a sand-heavy mix with 2 to 2.5 parts sand per 1 part clay. They bond well to the clay and straw in the LSC wall system, and are very user friendly in that they are easy to apply, not caustic to skin (just drying), very pleasurable to work with, easy to repair, and nontoxic. They also have a very long work time, making it easier to stop and start again on plaster projects.

Clay can be used from on site (ideal, as it has the least embodied energy and is most reflective of place) or purchased in bag form, usually in 50-pound bags. Ball and fire clays have great bonding and are much less dusty than kaolin clays. However, they are darker in color, so many people choose the kaolins for the final coat because you have brighter color options. (More on finishes in Chapter 9.)

Earthen plasters are excellent building skins. The one major downside of an earthen plaster is its water solubility. Water, especially wind-driven rain will slowly erode earthen plaster on the exterior of a building, creating a long-term maintenance issue. Earthen plasters also have a much lower compressive strength compared with lime plasters.

Lime Plasters

Lime plasters are simply plasters that contain lime as the binder. Lime is a caustic alkaline powder obtained by burning limestone. There are three types of lime: *hydrated* (Type S or Type N), *hydraulic* (natural or artificial) and *quicklime*. Hydrated lime is made by adding water to, or "slaking", quicklime (see below). Both hydrated and hydraulic lime set via a chemical change when water is added. Hydraulic lime will set underwater and is very similar to the cement used in concrete in that way. It might be the preferred plaster material for very wet climates or for deep repair jobs. It is sold in 50-pound bags. Natural hydraulic lime is currently distributed in North America only by the company St. Astier. It is very expensive, as it is imported from France. It comes in a variety of hardnesses, and you can get it ready-mixed (including lime and sand) called EcoMortar that allows you to simply empty the contents of the bag into the mixer and add a color packet (from the same company), and add water and mix.

Type-S hydrated lime can be found at masonry supply stores in 50-pound bags. (Type S is preferred over Type N lime for plaster because of its characteristics.) It ranges in price per bag from $7–14. It is more user friendly in that it has a much longer working time than hydraulic lime. It is also bright white and can be used with brighter pigments. It can be soaked in buckets ahead of time to theoretically activate the lime, although few plasterers maintain this practice.

Quicklime is a form of lime that reacts intensely when water is added to it, and is used to make hydrated lime. Quicklime must be "slaked" — soaked in water for years to increase its workability. This is traditionally the lime used in and on historic European buildings. For some plasterers, "slaked lime" that has been aged for years is prized for its workability. Quicklime is not recommended for most projects as it is hazardous to transport, and storing significant quantities of slaking lime requires a large amount of space.

Table 5.3: Environmental impacts of lime plaster

Ecosystem Impacts	Embodied Energy	Carbon Footprint	Indoor Environment	Waste
Moderate. Limestone is a nonrenewable resource but is abundantly available. Large scale quarrying can cause habitat destruction and surface and ground water interference and contamination.	**High.** 1.11 MJ/kg* or 116.9 MJ/m^2. Lime is processed at high temperature, in addition to quarrying and crushing energy input.	**High.** 0.174 $kgCO_2e$/kg* or 18.33 $kgCO_2e/m^2$. Lime will absorb CO_2 during the curing process, but due to fuel use during processing will still be a net carbon emitter, though accurate figures are difficult to assess.	**Very low.** Lime-based plasters can contribute to high indoor air quality, providing naturally antiseptic qualities and no toxic off gassing.	**Low to Moderate.** Construction: Plasters can be left in the environment or crushed to make aggregate. End of life: Plasters can be left in the environment or crushed to make aggregate. Embedded mesh will require separation.

Note:
* Data is from Inventory of Carbon and Energy (ICE) 2.0, University of Bath

Gypsum Plasters

Gypsum is similar to lime in that it is a dry powder that reacts with water to chemically harden. Gypsum plasters have a similar working time as lime plasters; however, they are much softer and remain water soluble, so are not suitable for exterior use.

Gypsum is purchased in 50-pound bags and is ready to mix with water and apply. Like lime plasters, gypsum plasters require a bit more skill than earthen plasters to get a quality finish.

In the United States, "synthetic" gypsum is a by-product of flue gas desulphurization (FGD) in the coal industry. "Of the FGD systems in the United States, 90 percent use limestone or lime as the sorbent. As the flue gas comes in contact with the slurry of calcium salts, sulfur dioxide (SO_2) reacts with the calcium to form hydrous calcium sulfate ($CaSO_4 \cdot 2H_2O$) or gypsum." (From the American Coal Ash Association website, www.acaa-usa.org.) A study conducted for the US government reported a significantly higher quantity of mercury in synthetic gypsum vs. natural: "Ten natural gypsum samples were compared to the twelve synthetic gypsum samples . . . The highest mercury concentration found in the natural gypsum was 0.03 μg/g compared to the lowest mercury concentration of synthetic gypsum of 0.10 μg/g." (USG Corporation, *Fate of Mercury in Synthetic Gypsum Used for Wallboard Production,* 2008.) Given the negative health and environmental impacts of coal and mercury, these findings are something to consider when thinking about using sheet rock or gypsum plaster. Making the choice even more difficult: there are no marketplace distinctions that identify a gypsum product as "natural" or "synthetic."

Table 5.4: Environmental impacts of gypsum sheathing

Ecosystem Impacts	Embodied Energy	Carbon Footprint	Indoor Environment	Waste
Moderate. Gypsum is a soft rock quarried from surface-based pits. Large scale quarrying can cause habitat destruction and surface and ground water interference and contamination.	**Moderate to High.** 6.75 MJ/kg* or 60.75 MJ/m^2 at ½ inch. Gypsum is processed using a moderate amount of heat. Fiberglass facing is applied, and may not be included in the above figure.	**Moderate.** 0.39 $kgCO_2e/kg$* or 3.51 $kgCO_2e/m^2$. Does not include production of fiberglass facing.	**Moderate.** *Exterior product:* Fiberglass particulate is shed during handling. The material may contain some quantity of toxic chemicals, including vinyl acetate monomer, acetaldehyde, and formaldehyde. Product would be used only on exterior of wall. *Interior product:* The paper and glue of interior drywall can be a good medium for mold growth in wet conditions.	**High.** Construction: Sheathing can be utilized strategically to minimize waste, but standard sizes can lead to high percentage of off cuts. Composite material cannot be composted or recycled. End of life: Cannot be recycled or composted. Will require separation from assembly.

Note:
* Data is from Inventory of Carbon and Energy (ICE) 2.0, University of Bath.

Sands

When it comes to plasters, be they earth or lime, the sand essentially performs as the bricks in a microscopic masonry wall. Good sand for plaster is clean (no salts, washed) and sharp, meaning it is comprised of sharp, broken particles — not rounded, soft particles like those found in beach sand. The sharpness aids in the structural integrity of the wall because those sharp edges lock together. Putting rounded sand in a plaster would be like trying to mortar together golf balls and expecting the result to stay solid under stress or movement. Sharp sand also resists dusting better than "softer" sands.

Sand comes in a variety of colors and particle sizes. It is very important to have a variety of particle sizes because homogeneity in plaster sand results in weak plaster; having a diversity of particle sizes adds strength. Bagged sand is usually rated by mesh size. Mesh size refers to the mesh used to sift the sand. The higher the mesh size number, the more holes in one square inch of mesh, and therefore the smaller the particle size. 100 mesh sand is very fine, whereas 30 mesh is very rough. A finish plaster should have a range of particle sizes from 30–100 or 60–100. Sand, being crushed rock, also comes in a variety of colors. It is best to use sand that will not diminish the color desired in the finish coat. Usually, a mason's sand is used in the first two coats of plaster and then a lighter silica sand is used in the final coat to keep the colors bright. Masonry sand can be ordered by the yard, whereas silica sands come in 100-pound bags.

Pigments

Pigments are ground-up minerals; they generally do not degrade over time. They are mixed into finish plasters to achieve a desired color. There are a handful of companies that specialize in natural pigments for natural plasters that are essentially nontoxic. It is possible to get pigments or mason stains from ceramic supply stores, although some of these are highly toxic minerals — not ideal for a house because they will inevitably "dust off" to some extent. Also, you have to be extremely careful not to inhale these pigments during the mixing process.

Some pigments are very potent and will stain most anything. Red iron oxide (oxidized iron — think blood, and how hard that is to get out of clothing) is a very potent nontoxic pigment. It will stain wood and clothes readily (pun intended); it is also one of three nontoxic iron oxides that can be used to make a wide range of earth tones. Yellow iron oxide and black iron oxide (actually a very deep blue) with red iron oxide provide all of the primary colors and a wonderful affordable and nontoxic color palette for a project. They can also be used to stain wood and concrete. See the photo of plaster samples in the color section of this book to get some idea of the color range possible with plasters.

Plaster Additives

Fibers

Fibers are commonly added to plasters to help alleviate cracking. Manures (cow or horse) have been used for centuries to add fibers and help with water resistance. Chopped straw (aka: non-animal processed cellulose) can also be added. Usually, there are bales left over after infilling the walls, and straw can be chopped in a variety of ways: with a sharp machete (for those wanting an upper body workout), a weed-whacker in a metal garbage bin, a lancelot (chainsaw attachment on a grinder, rather dangerous), a chainsaw, or a leaf-mulcher. Some people use a leaf blower in reverse as well, which sucks up the straw, chops it and deposits it conveniently in a bag. Other usable fibers include animal hair, plant fibers, and various synthetics. Pig hair was traditionally used in lime plasters because it doesn't break down in the caustic mix of lime and sand and is rather coarse. It is hard to acquire — as is a tolerance for the smell. Cattail fluff (the loose fibers from ripe cattail stalks in marshes) make a good non-visible natural fiber for finish plasters. Nylon fiber can be purchased by the pound and in a variety of lengths from masonry supply warehouses or ceramics suppliers.

Other Additives

There are a handful of very informative books on plastering that can give a much more detailed list of the various additives to plasters, and when, why, and where to use them. A list is in the Resources section.

Wheat paste is a common additive in "final plasters" (the last layer applied) to help control dusting and make a stickier mix. Wheat paste and sand is used as an adhesion coat over exposed studs in projects where using only natural materials is an objective.

Wheat paste is a mix of flour and water. Think papier-maché. It is very sticky and easy to make. Directions are covered in Chapter 9, Finishes.

Mica is a mineral added to finish plasters to give it a sparkling effect. Like anything in dust form, it is hazardous to breathe in, but it can make for a very attractive plaster. It can be found naturally in rocks or purchased from ceramics supply stores in a variety of mesh sizes.

Borax is also a plaster additive. It helps control dusting due to its soap-like quality. Borax helps align the clay particles and make the surface of a plaster tighter. I have experienced efflorescence with borax, but only with earthen plasters that used native soil. If using a native soil for plasters and planning to use borax in the mix, do a sizable test batch first.

Sheathing

There are a variety of materials to sheath an LSC building with, plywood being the most common. Generally, there is a sub-layer of plaster and an air space when using plywood for a rainscreen. However, in retrofits, remodels, or

Table 5.5: Environmental impacts of wood fiber sheathing

Ecosystem Impacts	Embodied Energy	Carbon Footprint	Indoor Environment	Waste
Low to Moderate. Most wood fiber products are made from post-industrial waste streams and do not directly involve the harvesting of timber. Third-party verified practices can be sourced, and practices should be confirmed with source to verify impacts. Most manufacturers use nontoxic binders; this should be confirmed.	**Moderate.** 9.36MJ/kg** or 71.3 MJ/m^2 at 1 inch. There are both wet and dry processing methods for wood fiber sheathing, and no third-party data is currently available to assess this category thoroughly. The additional energy efficiency added by these types of panels helps off set EE compared to plaster or non-insulating sheathing.	**Low.** There are no available figures for this product category. Particle board is rated at 0.86 $kgCO_2e/kg$* and uses similar processing techniques. Carbon sequestration potential for this material is high.	**Low to Moderate.** These products are made using a variety of binder materials. Many manufacturers advertise nontoxic binders, and some have third-party verification for low emissions. Sheathing is not in direct contact with indoor air.	**Low to High.** Construction: Sheathing can be utilized strategically to minimize waste, but standard sizes can lead to high percentage of off cuts. Wood waste can be recycled or composted. End of life: Can be recycled or composted. Will require separation from assembly.

Note:
* Data is from Inventory of Carbon and Energy (ICE) 2.0, University of Bath
** From Environmental Building News, Assessing Sheathing Options

non-inhabited spaces, plywood has been used directly up against the LSC.

For interiors, there are a wide variety of surface treatments: lath and plaster, sheet rock, tongue-and-groove wood, and a variety of decorative sheet goods.

Given that plywood is used in nearly every project for some element — roof sheathing, cabinetry, floor sheathing, strongback forms, etc. — it is important to have a grasp on the material's benefits and disadvantages.

Wood

Humans have been working with wood since we fashioned the first spears out of wood over 400,000 years ago. Our hands and minds have evolved working with this highly variable, versatile, and extremely useful material. From the Neolithic huts of Europe to the exquisite craftsmanship of Japanese tea houses, wood has been used for everything from fuel to fiber, and, sadly, it has been greatly overused. The wise and ethical use of wood is paramount if we are to preserve and restore our precious forests worldwide. To design an ecological house is to take the health of our forest ecosystems into account, and, as the sustainable wood harvesting industry grows, it is important to make wise choices as homeowners, builders, and designers.

Building with wood sourced as close as possible to the property you are building on, using local mills, and simply designing to minimize the use of wood are good steps toward conserving this precious, yet renewable, resource. Using small-diameter trees in place of old growth and its sought-after CVG (clear vertical grain) and reusing or repurposing wood from other projects are two other ways of having a positive impact while using wood as a building material.

Wood and straw have a very similar chemical composition and therefore are very similar in their carbon sequestration potential. However, straw is an annual crop, whereas (depending on the species) wood may require centuries to reach reproductive maturity. Douglas fir, often used for building on the West Coast, reaches its reproductive peak at 250 years of age, yet depending on the climate, it is harvested every 30–80 years.

There are a handful of organizations that certify wood as being sustainably harvested. The most commonly seen is the Forest Stewardship

Table 5.6: Environmental impacts of wood framing

Ecosystem Impacts	Embodied Energy	Carbon Footprint	Indoor Environment	Waste
Low to High. Forestry practices can range from third-party verified sustainable harvesting to unregulated clear cutting. Confirm practices with source to verify degree of impact.	**Low.** 7.4 MJ/kg* (spruce lumber) or 43.66 MJ per 2x4x8. Quantities of lumber used for different prefabricated wall systems will vary widely, and total embodied energy figures must be assessed based on design.	**Low.** 0.59 $kgCO_2e/kg$.* Quantities of lumber used for different wall systems will vary widely, and total carbon footprint must be assessed based on design. High carbon sequestration potential.	**Low.** Framing lumber in most panel systems is not in direct contact with indoor air, but softwood lumber does not have toxic off gassing or contain any red list chemicals.	**Low to High.** Construction: Framing lumber can be utilized strategically to minimize waste, but standard lengths can lead to high percentage of off cuts. Wood waste can be recycled or composted. End of life: Can be recycled or composted. Will require separation from assembly.

Note:
* Data is from Inventory of Carbon and Energy (ICE) 2.0, University of Bath

Council (FSC), a global entity. The FSC certifies companies based on their having met ten principles of sustainability in their harvesting and processing methods. It is common to go to a big box store like Home Depot or Lowe's and find dimensional lumber or lumber products that bear the FSC stamp. This is a step in the right direction for acquiring sustainably managed wood. Keep in mind, though, that the larger a company is, the less place-based its management is likely to be, which might lead it to compromise ideals for a bottom line. The best bet is to always strive to get as many building materials from as close to the building site as possible.

One of the bonuses of an LSC wall system is that it works well with just about any wooden wall framing system. From a materials perspective, being able to space wood members farther apart allows you to use less wood, which is a good thing for the environment; in addition, wide spacing saves time and labor.

People

Although you might be the type who enjoys the challenges of solo work, given that some of the work is pretty tedious, and that one method of installation requires a *lack* of a roof, making this a multi-worker project is a much better option. You will usually need a minimum of five people — for prepping and moving materials, mixing, and tamping. Work parties are an excellent way to procure some extra help and to have good clean family fun together. But there are many ways to source help beyond your friends and family (who may or may not be into your crazy idea of building a house with walls of straw and clay). See Chapter 8 for more on workshops and how to host a good one.

Chapter 6

Design Options: Framing Systems and Form Options

THERE ARE A VARIETY OF WAYS TO CREATE a superstructure to infill with light straw clay. The framing systems we will explore here are the following: timber frame, Larsen truss frame, pole frame, split studs, LVL, conventional, and retrofits. Each has its benefits and challenges. We will mainly explore the details necessary to know for a light straw clay infill and not necessarily focus on a step-by-step of how to frame a structure.

Post and Beam

A very simple way to frame for a LSC wall system is post-and-beam construction. Instead of the conventional framing based around 4×8 sheet goods and 16–24″ spacing for batt insulation, post and beam can create a superstructure using slightly larger posts for the superstructure. Then there is a variety of choices for the secondary framing that will hold the LSC.

Post-and-beam framing is commonly used with straw bale construction to do away with the need for the excessive notching needed to imbed straw bales into a traditional 16″ on-center framing systems. It requires less wood, and any carpenter can do it because the metal fasteners and brackets used to hold the posts and beams together do not require a high level of skill to assemble (unlike traditional timber framing, which is much more difficult to do right). These wall systems can also be easily engineered and can accommodate a variety of shear systems. Drawings for timber frame walls can easily be applied to post-and-beam construction.

LVL Framing

A similar type of post-and-beam construction is framing with LVLs (laminated veneer lumber). LVLs are like industrial-strength framing members that are essentially very robust plywood. Due to their engineered strength, they can span long distances, making for a superstructure with vertical members that can be spaced at 8–10′ intervals. It is much like timber framing in spacing, but glue and smaller-diameter trees are used to achieve the same strength. A secondary framing system is still needed between the structural spans.

Advantages:

- Very quick to go up.
- Very straight materials are used (no bowing or oddities).
- Can be easily engineered.
- Compatible with conventional framer's skill set.
- Very strong.

Disadvantages:

- LVLs are expensive.
- The glues used are likely formaldehyde-based.
- Framing members are very heavy.

This framing system works best with a box beam assembly, where a simple box is built as a top plate instead of notching in another horizontal LVL to carry the load of the roof or second floor.

Timber Framing

Timber framing is a style of building that involves mortise and tenon joinery and no metal fasteners. It is both an art and a study in physics. It relies on the strength and the beauty of larger timbers, and the structure of a building is exposed and celebrated. The timber frame buildings of Europe are where light straw clay got its start. In lieu of the heavy clay daub that was traditionally used to infill between timbers, light straw clay was used instead, creating a lighter load to carry up to second and third floors (and also increase the insulation value of the walls).

Timber framing requires good planning, good wood, excellent skills, and good tools. Traditionally done with hand tools, contemporary builders implement a wide variety of tools. Precision is key, as part of the strength in a timber frame is the tightness and accuracy of the joints.

Timber frame building with lime plaster and LSC wall system.
PHOTO CREDIT: LYDIA DOLEMAN

Traditional timber frame with no metal fasteners.
PHOTO CREDIT: LYDIA DOLEMAN

Advantages:

- Beautiful and elegant; wood/structure is exposed and not hidden in the wall (although it can be).
- Infill walls can be in the same plane, set to the inside or outside, or completely independent of the timber frame superstructure.
- Infill walls can be framed out of smaller pieces of wood, as they are nonstructural.

Disadvantages:

- Potentially expensive.
- Requires a high level of skill.
- Difficult to engineer a "true" (no metal fasteners) timber frame in a seismic zone.
- Requires large timbers (aka: older trees from an established forest).
- Green timbers can shrink over time, leaving cracks where wood meets plaster.

The most basic way of incorporating light straw clay into a timber frame is to infill with some secondary framing (Larsen truss, split stud, or conventional framing) between the vertical posts. Important details to consider are how much of the timber frame to leave exposed and how thick the anticipated coats of plaster will be. To ensure a tight building envelope, you'll need to plan for shrinkage — of the wood over time and of the light straw clay as it dries. Appropriate air-fin installation has to take place prior to the installation of the secondary frame and the light straw clay infill. Another consideration is the roof overhang if your timber frame is set to the inside or to the outside of the building.

PHOTO CREDIT: LYDIA DOLEMAN

PHOTO CREDIT: ALLISAN BUCKINGHAM

PHOTO CREDIT: LYDIA DOLEMAN

Cross-section of a Timber Frame Structure

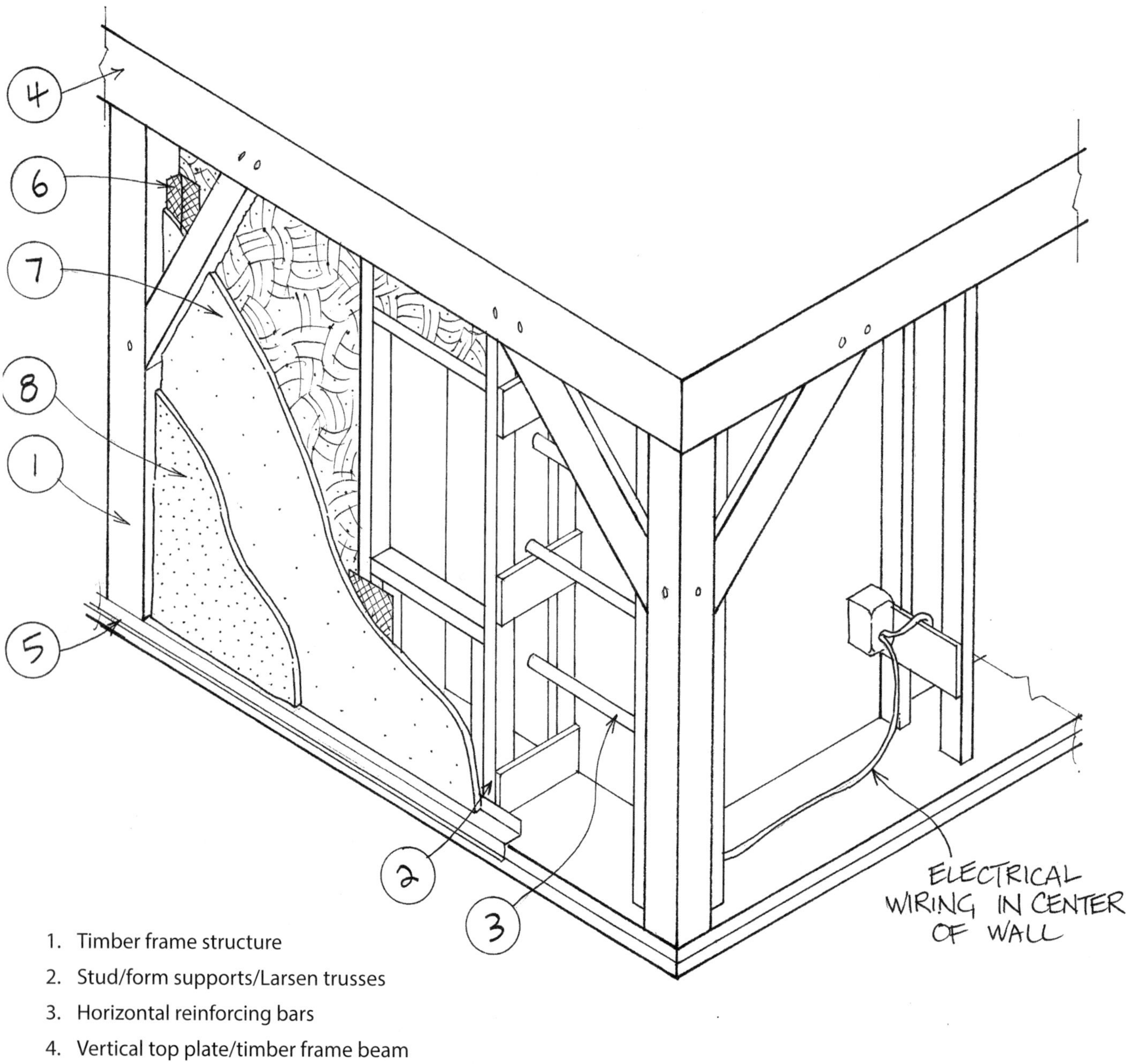

1. Timber frame structure
2. Stud/form supports/Larsen trusses
3. Horizontal reinforcing bars
4. Vertical top plate/timber frame beam
5. Flashing at bottom of wall
6. Mesh or burlap
7. Rough coat of plaster
8. Brown coat of plaster

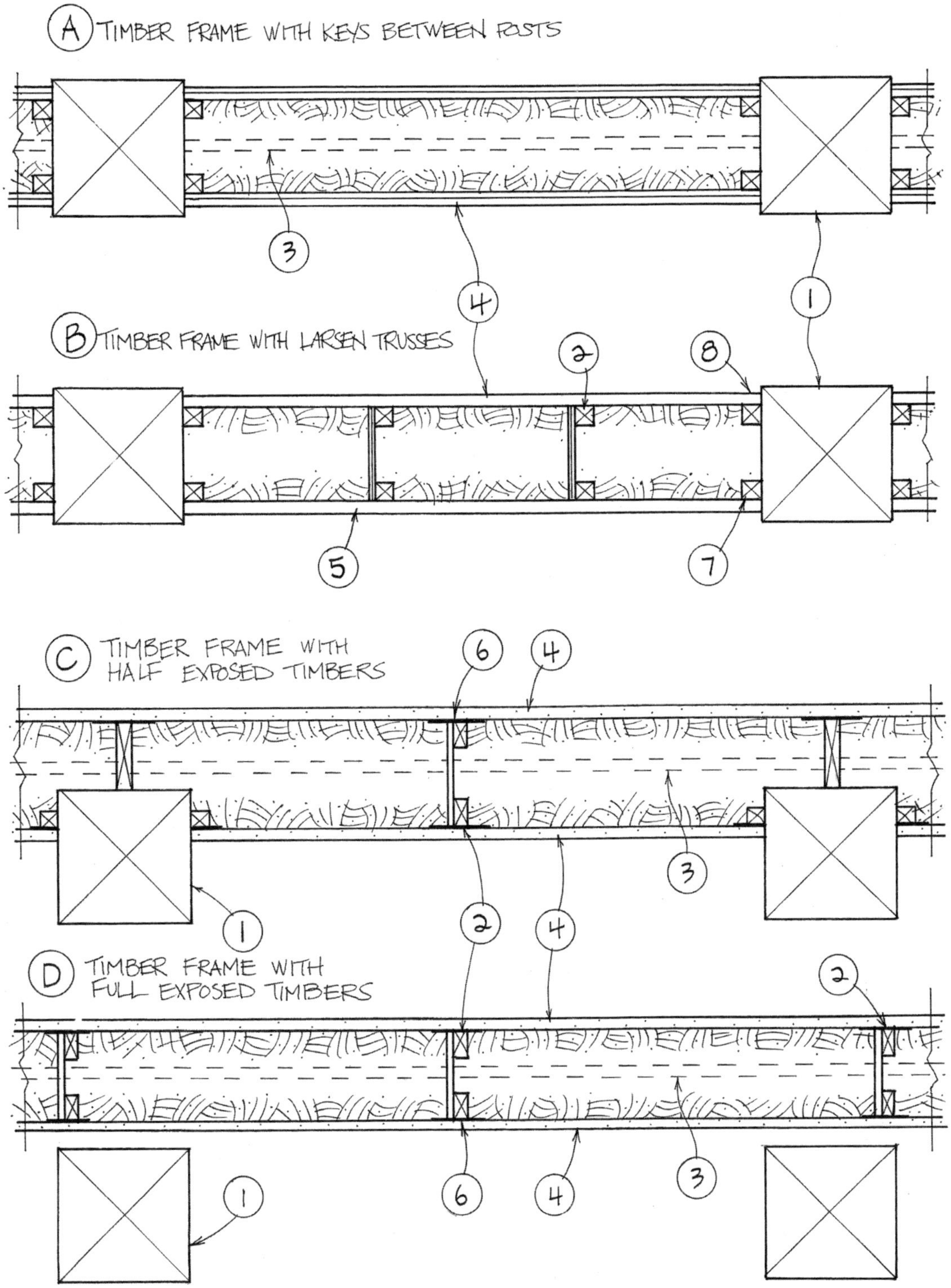

1. Timber frame structure
2. Stud/form supports/Larsen trusses
3. Horizontal reinforcing bars
4. Plaster skin
5. Three-coat plaster skin
6. Mesh or burlap
7. Air fin
8. Air fin with J-channel

Larsen Truss System

The Larsen truss system is a wall framing system that incorporates a ladder-like stud in place of a much wider stud. Typically placed on 24″ centers (though spacing may range from 16–48″ centers, depending on design loads), these custom-made studs can be bulk manufactured on site out of scrap wood and can create whatever wall thickness is required. They also allow for a continuous thermal envelope by allowing for largely uninterrupted light straw clay insulation. This system is also convenient when it comes to running electrical wires because they can be run inside the cavity, thus reducing the need to drill through studs.

Typically, two 2×2 are laid in a form and stapled, screwed or nailed (pneumatic staples are the fastest) with cross pieces of wood or plywood that has been cut to a predetermined, uniform length. For a wall thickness of 12″, the cross pieces, or rungs of the ladder, would be cut to 12″ and placed at various strategic heights. If you know you are putting your electrical boxes at particular heights, you can choose to put the cross pieces at different heights to make it easier when you mount the boxes. Also, the uniformity in the horizontal pieces makes laying any horizontal reinforcement easier, as some wall systems benefit from the installation of these to help strengthen a wall from out-of-plane loads.

Larsen trusses are used to mimic 2×6 framing for a continuous thermal envelope. PHOTO CREDIT: LYDIA DOLEMAN

Larsen trusses can be made out of a variety of dimensional lumber, however, anything thinner than 2× material (like your standard 2×4 or 2×6) becomes challenging to fasten formwork to. Not an impossibility to use 1× (half the thickness of the above), but something to consider.

The Larsen trusses are fastened to a top and bottom plate, much like conventional construction. In some instances, where there is a superstructure, it is possible to notch a 2x top plate on edge into the outer studs or not to incorporate a top plate at all, leaving the top portion of the wall very easy to fill up to the soffit or roof intersection.

Advantages:

- Use of small dimensional wood.
- Use of scrap wood.
- Very easy to construct.
- Compatible with all superstructure framing systems.
- Continuous thermal envelope.
- Easy for electrical wiring and plumbing to be routed through walls.
- Lightweight.
- Modular.

Disadvantages:

- Requires labor to construct trusses.
- Less wood for formwork to be attached to.
- 2x2s can be a challenge to firmly attach to bottom plates without blowing out/splitting wood.

Corner Options for Larsen Truss Framing

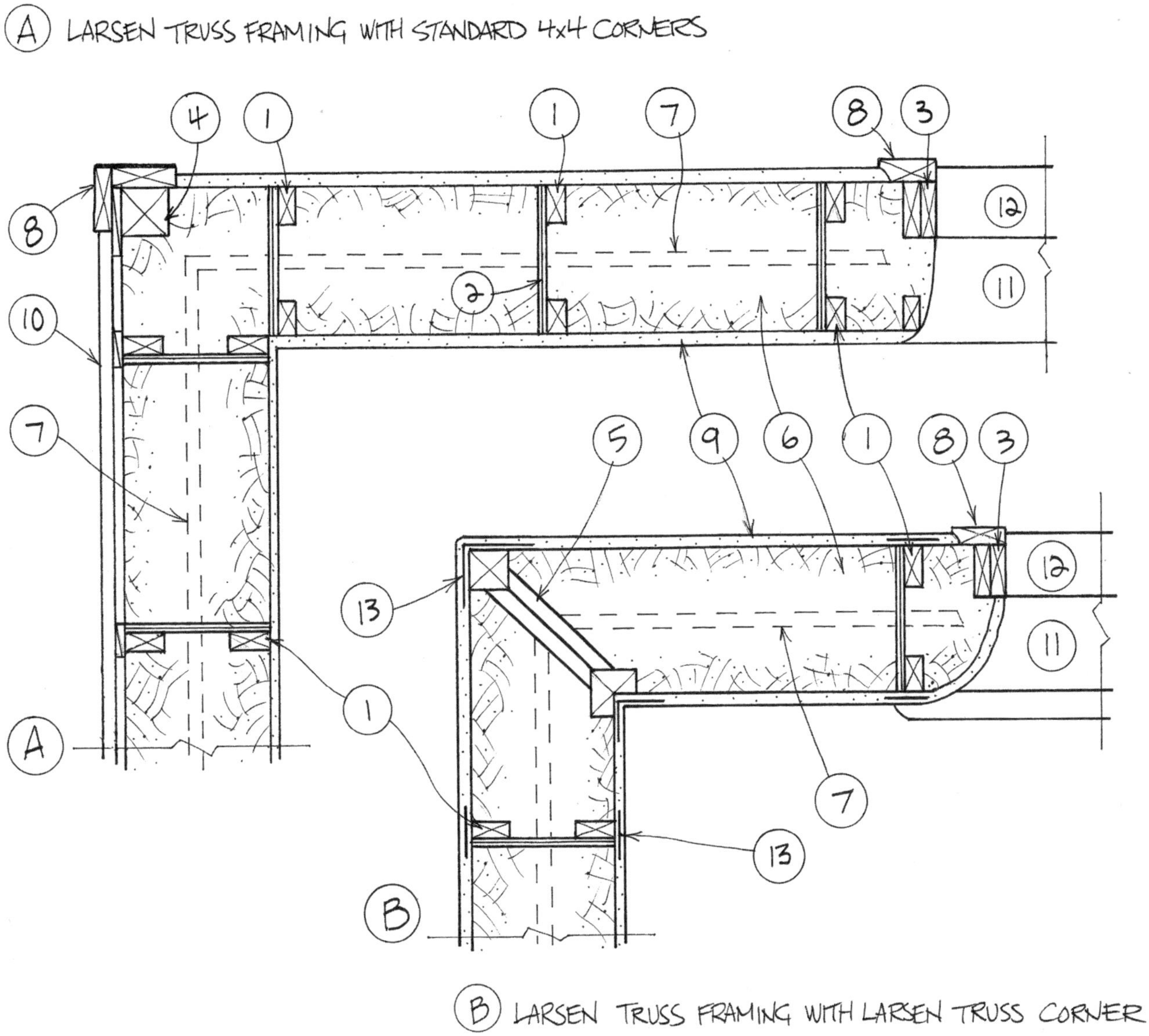

1. Larsen truss
2. Gusset
3. Window framing
4. 4×4
5. Larsen truss corner post
6. LSC
7. Horizontal reinforcing bar
8. Trim
9. Plaster
10. Rainscreen
11. Sill
12. Window
13. Paper and then burlap or mesh

Double Stud or Split Stud Framing

Larsen trusses are very similar to double stud and/or split stud framing. Double stud framing is an easy transition from conventional framing in that it is simply two conventionally framed walls spaced the desired distance apart. It can also be used without an auxiliary superstructure, like post and beam or timber frame. For a 12″ wall thickness, 2×4 studs are placed on 24″ centers, wrapping the exterior of the building, then a second wall is framed on the interior using 2×4 studs, with bottom and top plates placed 12″ exterior to interior apart. This can also be framed using a 2×12 top and bottom plate.

Generally, split stud construction staggers the outer wall framing from the interior wall framing; typically, this is done with 2×4 studs placed flat for more insulation between, or with the 1½″ face out in conventional stud framing style. For taller wall sections, it may be beneficial to frame with the 1½″ face out to prevent the wall from bowing or to keep studs in line with one another; essentially, you make partial Larsen trusses to prevent bowing. Care needs to be taken to plane out or remedy any bowing of the framing, as that will be a big headache that is just transferred to the plasterers or carpenters.

Split Stud Framing with Timber Frame

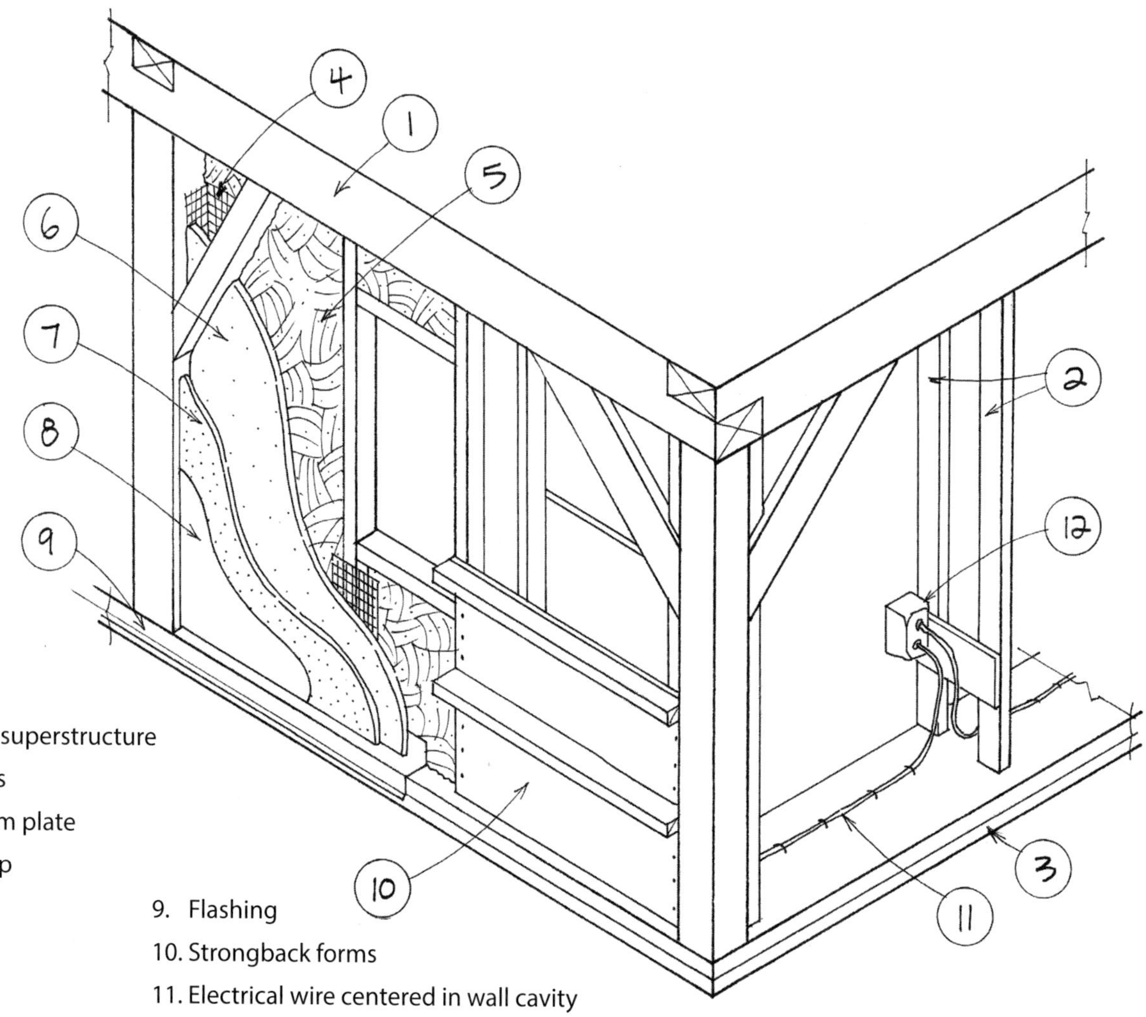

1. Timber frame superstructure
2. 2×4 split studs
3. Double bottom plate
4. Mesh or burlap
5. LSC
6. Rough coat
7. Brown coat
8. Finish plaster
9. Flashing
10. Strongback forms
11. Electrical wire centered in wall cavity
12. Electrical box mounted flush to studs

Advantages:

- Very straightforward for conventional framers.
- Continuous thermal envelope.
- More room for attaching formwork.
- Labor savings in not making trusses.
- Possible to mount windows to framing.

Disadvantages:

- More wood use than single framing.
- Flat framing can result in bowed walls if precautions not taken.

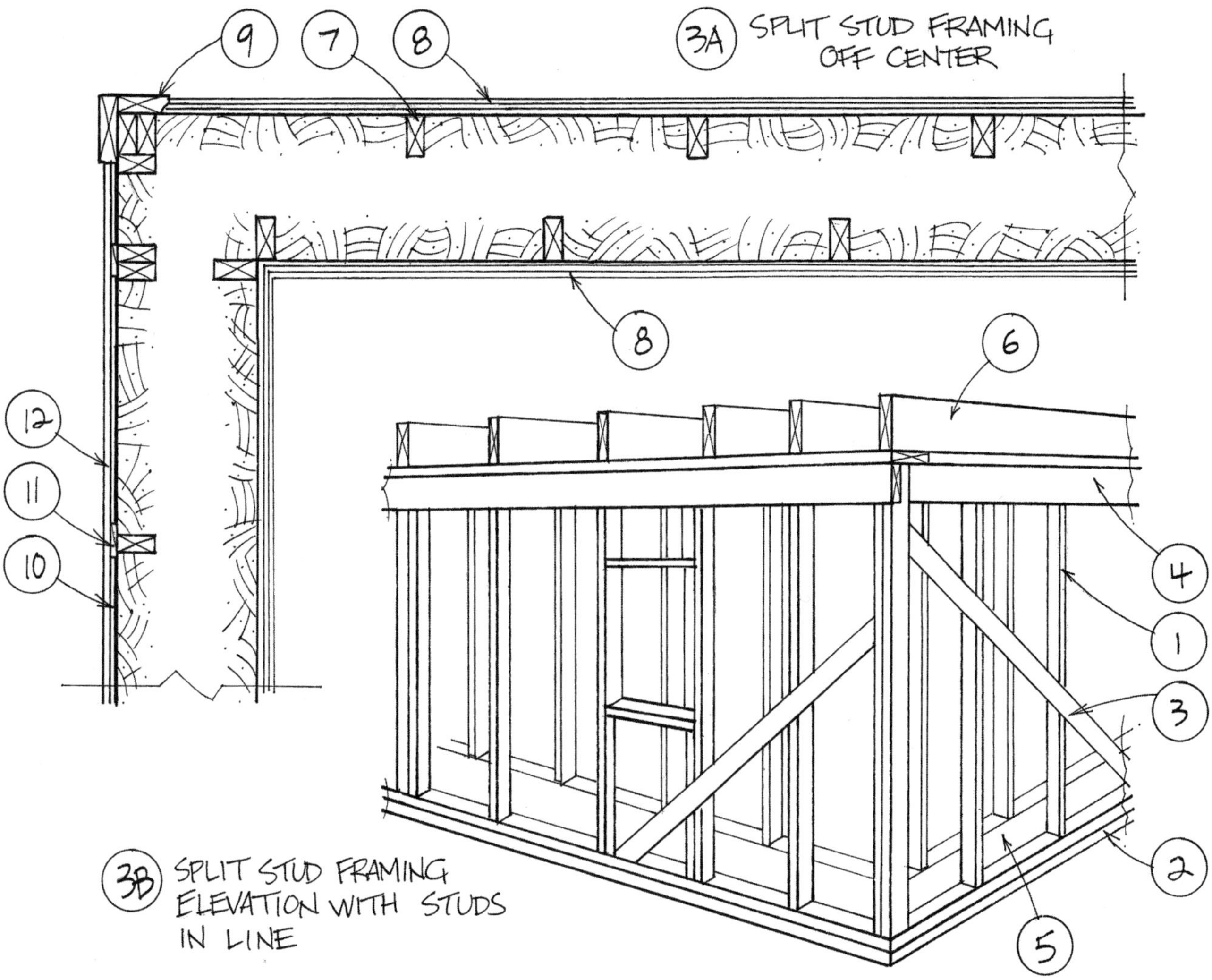

1. Studs
2. Double bottom plate
3. Let-in cross bracing
4. Vertical top plate
5. Insulation between bottom plate rails
6. Joists
7. Offset studs
8. Coat plaster skin
9. Exterior corner trim
10. Single coat of plaster
11. Furring strip
12. Rainscreen

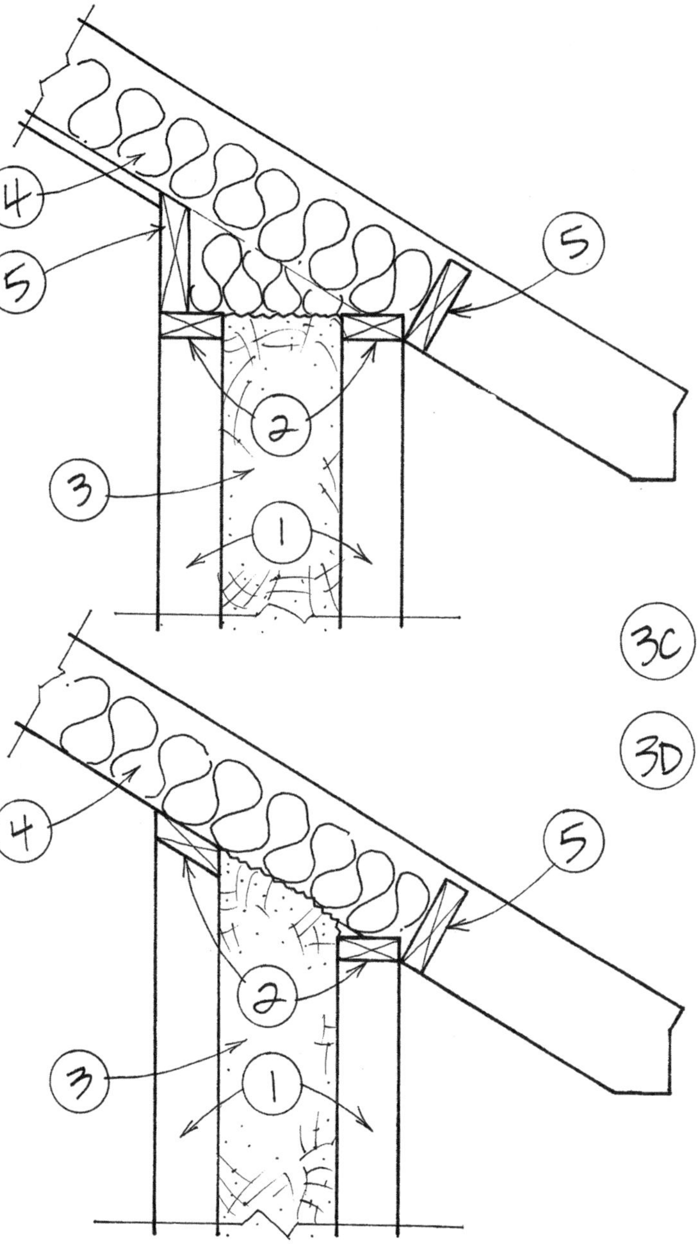

3C SPLIT STUDS WITH
3D SEPARATE PLATES

1. Studs
2. Top plates
3. LSC
4. Ceiling insulation
5. Blocking

Pole Frame Building

Pole frame building is a traditional style of framing that uses round poles from tree trunks that have not been milled into dimensional lumber. Poles are best cut in the spring and peeled while the sap is flowing and then set aside to dry for a few weeks before being used for construction. Ideally, they are stacked or stored in a manner that encourages airflow, so the wet, freshly peeled poles aren't encouraged to get mold spots.

In some situations, it is highly beneficial to harvest some of the competing understory trees in a forest where you have a wide age range. Understory trees that are 4–10″ in diameter are likely stressed because they are growing under the canopy of larger, taller trees. These understory trees are called "suppressed growth," and they tend to have tight growth rings and higher strength. In a forest of similarly aged trees, removing some allows more resources for the selected remaining trees.

Pole framing can be done in two ways: with mortise and tenon joinery, or lag bolted together.

Given the nature of working with round wood, it is extremely difficult and time consuming to construct a pole frame structure without metal fasteners. Most pole frame buildings are put together using lag bolts, timber lock screws, or some equivalent. Just as timber framing can be aesthetically very rustic or utilitarian, so can pole framing. This is a framing system that some people specialize in. If you are interested in framing a whole house this way, I highly recommended that you either start with a small project yourself or hire someone with experience. Working with rounded, tapered framing materials requires a lot of attention to many details.

Localized materials at their finest.

PHOTO CREDIT: DEAN HAWN

Note the curves in these peeled poles. In the forest, they looked absolutely straight! It's part of the challenges and beauties of pole framing.

PHOTO CREDIT: LYDIA DOLEMAN

Straw clay contours to the variations in pole shapes perfectly.

PHOTO CREDIT: LYDIA DOLEMAN

Poles can be very challenging; essentially, poles are elongated cones (not tubes) which makes the fundamentals of good carpentry — plumb, level, and square — very hard to achieve. However, the flexibility of LSC works perfectly with pole-framed structures, and a plaster skin can beautifully highlight the organic qualities of pole frames.

Large poles can be the superstructure, and smaller poles can be used in a split stud-style of framing or to make Larsen trusses. It is possible to purchase poles that have been debarked and essentially lathed to be nearly perfect cylinders. These could make pole framing much more straightforward by eliminating the taper, but would add cost (and may not even be available in your area).

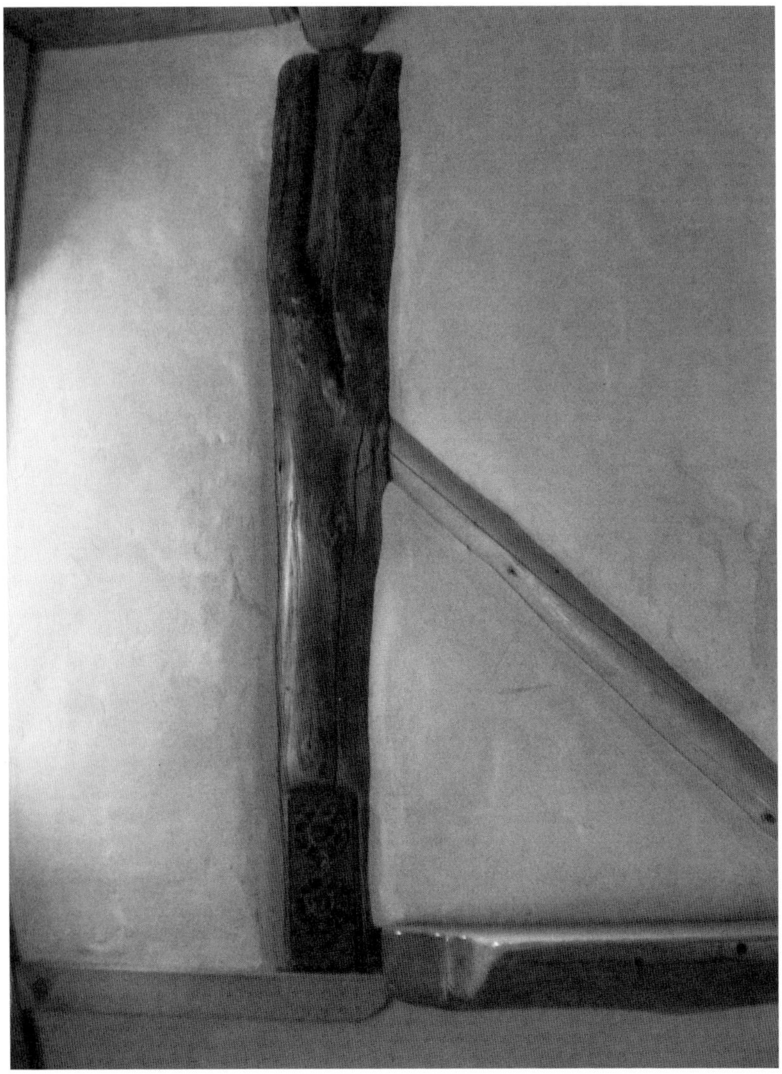

Plaster can highlight the beauty of non-linear wood.
PHOTO CREDIT: LYDIA DOLEMAN

Advantages:

- Use of wood from on site.
- Minimal embodied energy in the wood.
- Round wood is significantly stronger than sawn lumber.
- Straw clay and plaster are just about the best wall systems to encapsulate and/or highlight a pole frame structure due to their non-linear nature.
- Can be a very place-based style of building.

Disadvantages:

- Time of year for harvest may not sync with building schedule.
- Poles are rarely perfectly straight, and all have a slight taper, making plumb, level, and square construction a unique challenge.
- Detailing a pole frame structure is challenging.
- Forms up against poles have to be custom cut at times.
- Finding experienced pole framers or information on pole framing may be difficult.
- Engineering a pole-framed building requires expertise.
- Some types of wood are not ideal for carrying loads horizontally.

Notes on framing in general: when in doubt, hire out! Read building plans thoroughly. Talk to tradespeople during the design phase to work out details on paper. If doing it yourself, start with a small project first.

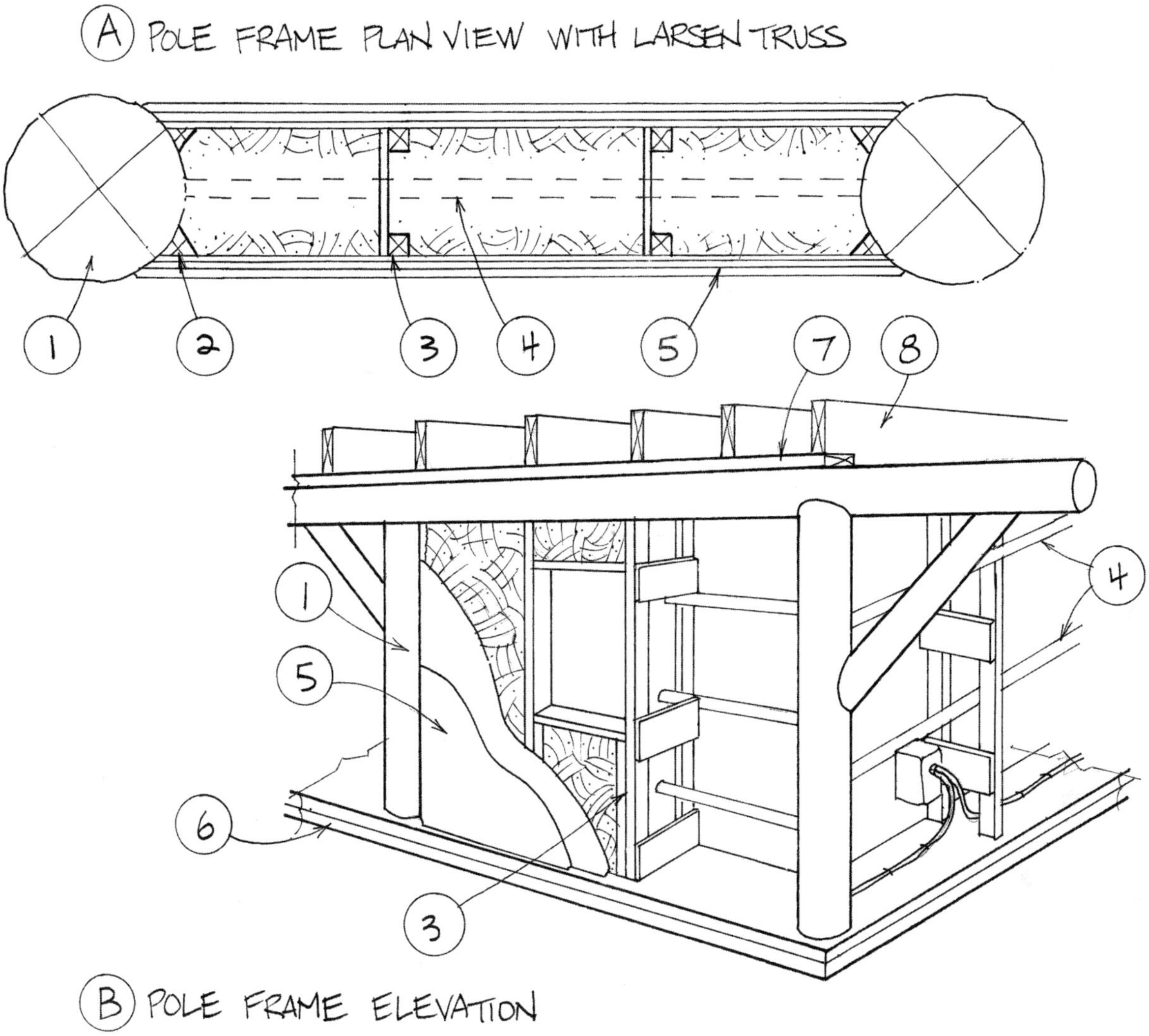

1. Vertical pole superstructure
2. Key and form support
3. Larsen truss
4. Horizontal reinforcements
5. Plaster skins
6. Double bottom plate
7. Leveling plate for joists
8. Joists

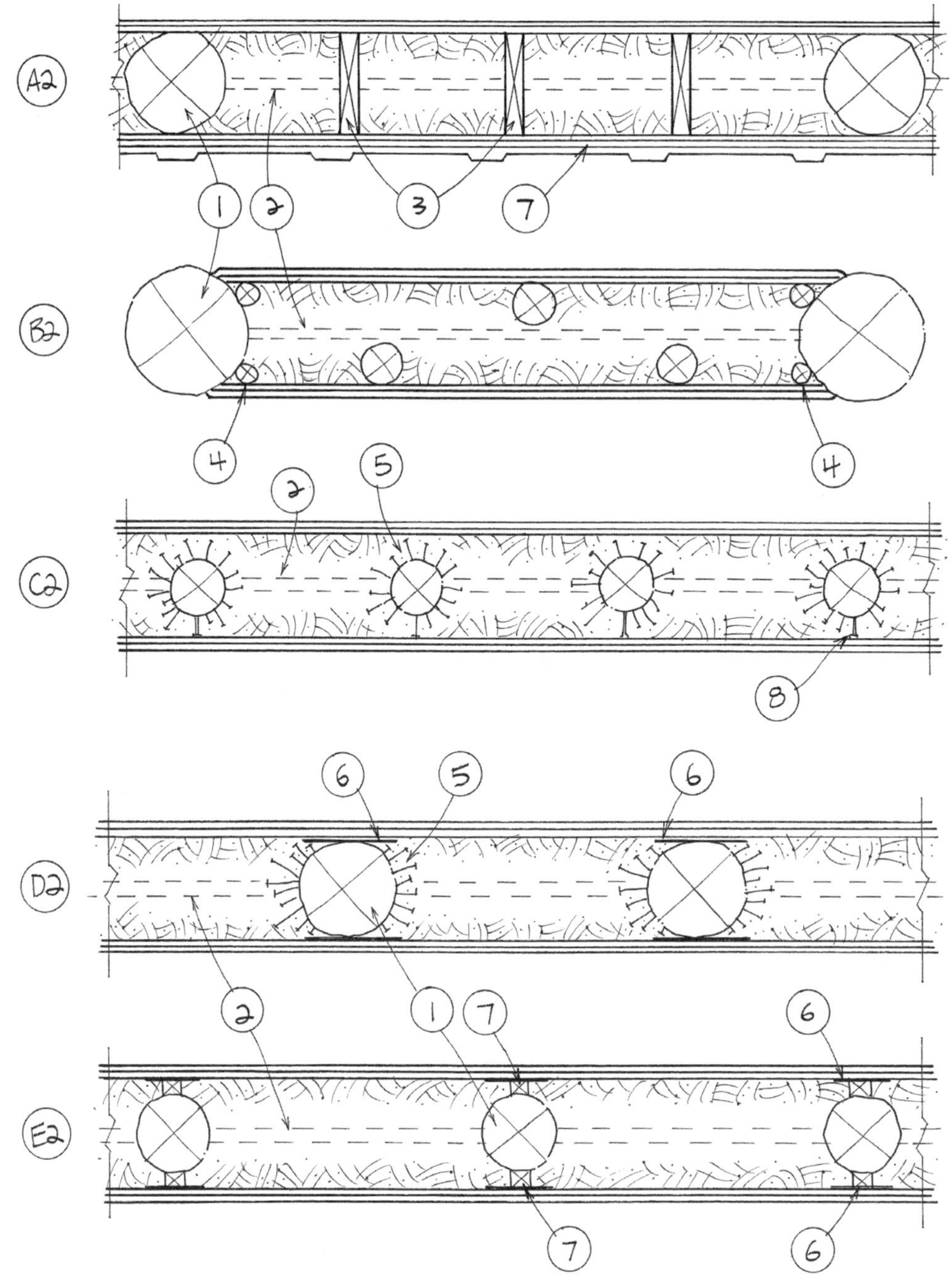

(A2) PLAN WITH STUDS

(B2) PLAN WITH SECONDARY POLES

(C2) PLAN WITH CONTINUOUS ENVELOPE AND BURIED POLES

(D2) PLAN WITH BURIED POLES

(E2) PLAN WITH BURIED POLES AND MOUNTING STRIPS

1. Pole superstructure
2. Horizontal reinforcing bars
3. Studs
4. Secondary poles for key and form support
5. Nails to hold/key LSC to poles
6. Paper and burlap
7. Furring strips
8. Fastener made of nails or screws to hold forms out

Retrofits

Here is a great opportunity to use light straw clay. This topic could really be a book in its own right. Remodeling an existing structure is always like opening a can of worms or pulling the loose string on a knitted sweater, but if you have a superstructure that is undergoing a major remodel, LSC is a great natural wall system to consider. One great advantage to it is that you can decide to either keep the existing wall thickness or fur out with Larsen trusses or split stud construction to increase the wall thickness and your insulating capacity. If the exterior is already sided, it is possible to install a permanent formwork on the interior and infill with a very light LSC mix and then dry the building on the inside.

LSC can easily be used in interior wall applications, adding a sound-dampening quality to interior walls. One could even cover an LSC wall system with drywall and have it seamlessly match an existing structure. However, latex paints are not recommended over any natural wall system because they act as a vapor barrier, trapping moisture in a wall, which is likely to cause problems.

For complete remodels or retrofits, a rainscreen or exterior plaster skin should be applied, and care should be taken to maintain the shear strength of a wall that has had its exterior siding removed. This can be achieved with metal cross bracing, "let-in" diagonal bracing, or control panels with plywood on corners. If you are in a seismic area, consulting an engineer is key to maintaining the structural integrity of a building that once had siding or plywood on the exterior. Shear walls can be created on the interior of a building to avoid having to add permanent plywood to the exterior; shear walls on the interior help maintain breathability and a homogenous plane on the exterior of a LSC retrofit/remodel.

This retrofit has LSC applied separately to the interior and the exterior of the building. An extension of the foundation was built to carry the exterior loads. Note the metal cross bracing and the framing for a rounded exterior corner.
PHOTO CREDIT: MARK LAKEMAN

When infilling with light straw clay in conventionally framed structures (i.e. not split stud frames, but standard single stud framing), it is beneficial to create "keys" in the studs to help limit shrinkage as the LSC dries. This will prevent drafts as well as stabilizing the panels, which are essentially floating in the bays.

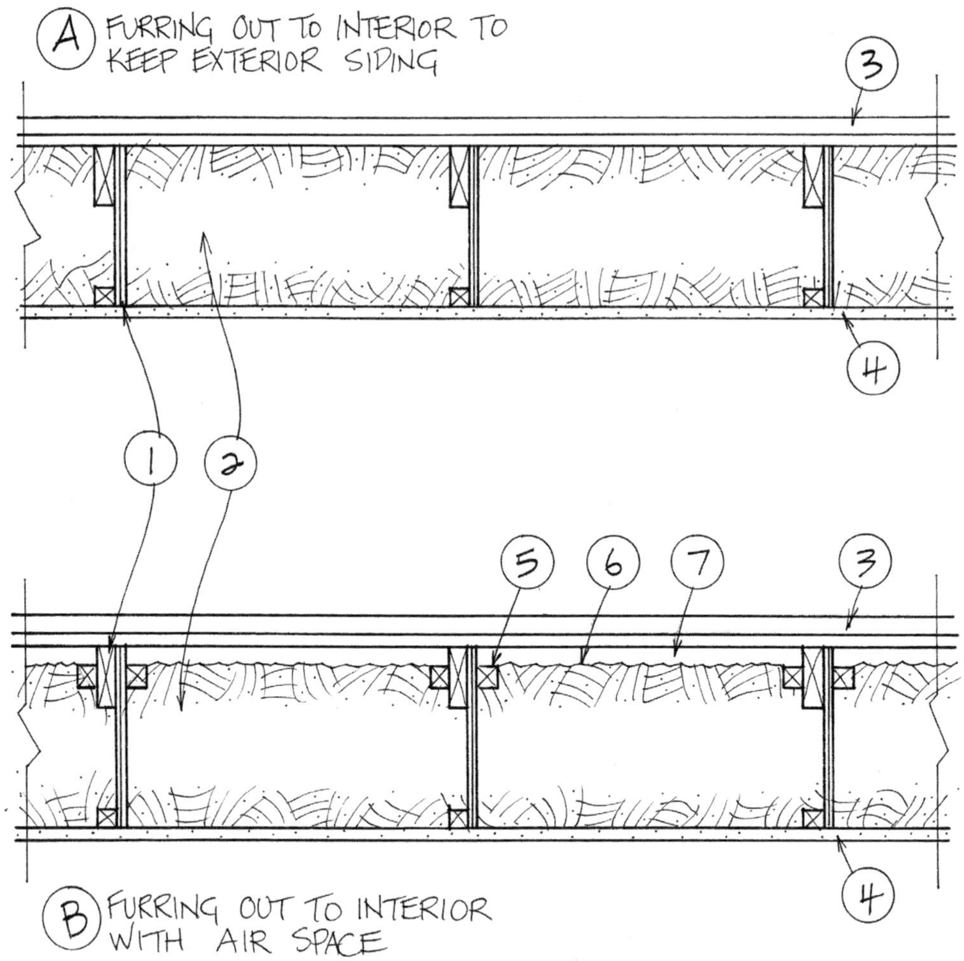

1. Larsen truss
2. LSC
3. Exterior siding over sheathing
4. Plaster skin
5. 1×1 or 2×2 to hold mesh
6. Mesh
7. Airspace

Keys are tiny additions of wood to a frame that increases the surface area to help mitigate shrinkage gaps. In addition to this, they help break up direct drafts by both reducing the amount of shrinkage between the LSC and the framing, and by creating a little "jog" for the air to flow around, thereby reducing drafts.

For conventional framing, horizontal stabilizing bars are essential if there is to be plaster on both sides. The stabilizing bars can be made of wood, bamboo, wire, poly twine — anything that will create an internal skeleton within the stud bay to help support the light straw clay after it dries. This adds some support to the free-floating panels. Air fins are also important for this type of LSC, as there is not a continuous thermal envelope.

Challenges with Retrofits

- **Relocating electrical**
  - This is encountered in most remodeling projects. Rerouting lines, running conduit and relocating outlets and switches can all be a part of a retrofit/remodel.
- **Drying walls appropriately**
  - Using fans and dehumidifiers is very important in drying out LSC walls where one side is closed off to ventilation.

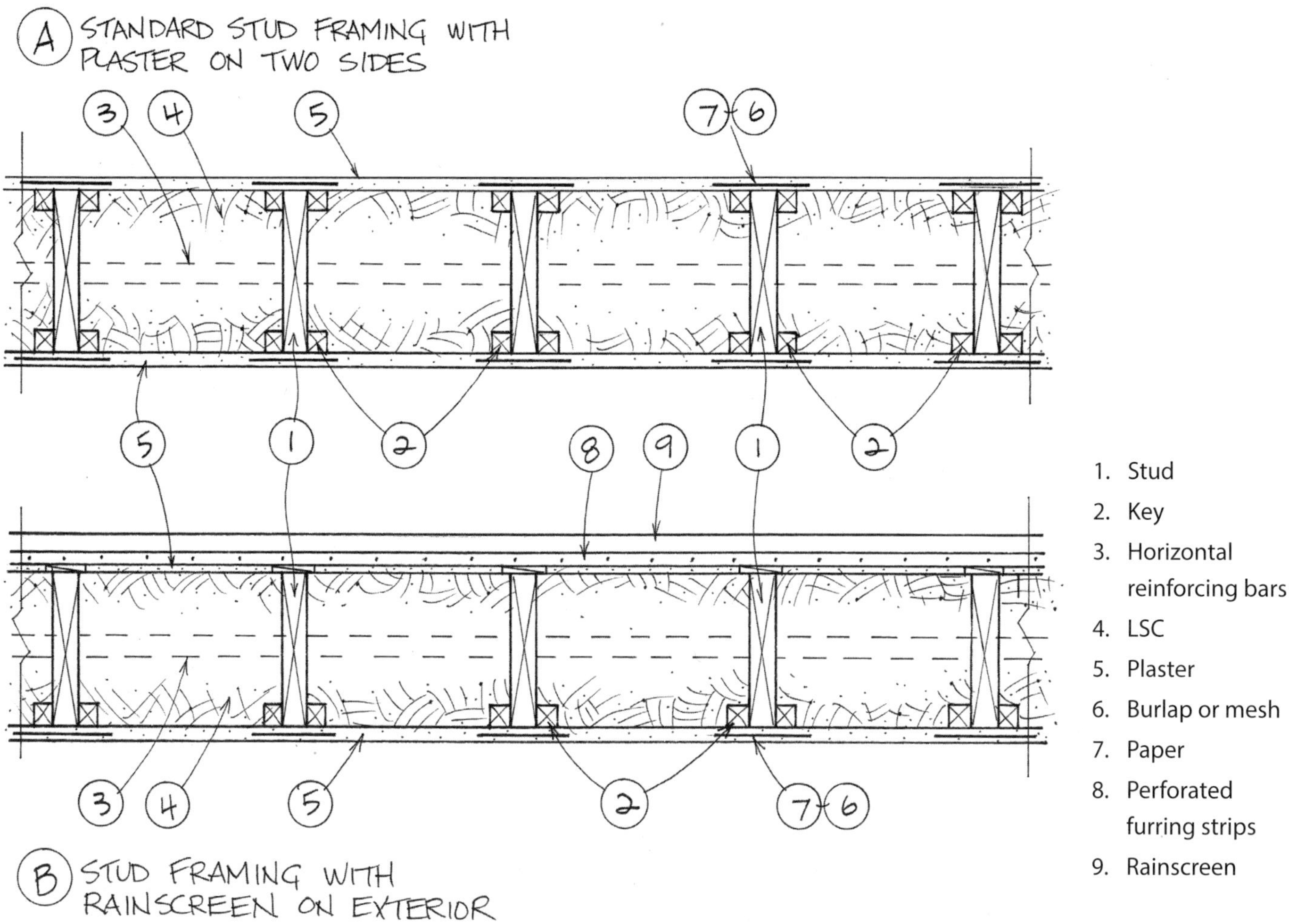

- Removing siding while building dries and then replacing it may be a viable option.
- Perforating the LSC to encourage moisture to exit can expedite the drying process.
- Use a very light mix if using permanent formwork.
- Regularly probe with a moisture meter to monitor moisture levels.

- **Preexisting vapor barriers on the exterior**
  - This can be handled by removing exterior siding and furring out for a rainscreen.
- **Encountering preexisting mold issues**
  - Pinpoint the source and repair it.
  - Apply hydrogen peroxide to kill mold and spores.
  - Allow time for affected area to adequately dry out.
  - Monitor the area.
- **Structural issues with the increase in wall weight and potential engineering issues**
  - Talk with an engineer prior to starting work.
  - Reinforce floor joists, if necessary.

Formwork

Formwork is essentially a mold for retaining and shaping the insulation. There are two types of formwork in a light straw clay structure: temporary, and permanent.

Temporary formwork

Temporary formwork generally consists of plywood (or planks) that is screwed to studs or trusses and is then removed after installation of the LSC. The most basic formwork is just a section of plywood. The thicker it is, the less likely it is to bow out, and the more likely you are to achieve a flat wall (which is essential for a wall that will have a rainscreen or cabinetry). For consistently flat walls, forms that implement the "strongback" system are preferred. The strongback system involves a 2×4 or 2×6 fastened to the top edge of the form. It is overhanging just enough to lock in the form that goes above it, so the bottom of the subsequent form is held tight against the wall.

Strongbacks leapfrog and hold each other in place. Photo credit: Jim Rieland

Any screw that is long enough to go through the formwork and then into the structural support a minimum of ¾″ is sufficient. The rule of thumb is that the fastener should penetrate two times the thickness of the form (so, ¾″ plywood would require a 2¼″ screw). However, the constant reuse of a screw can cause it to strip out, break, or become difficult to find. Using hex head roofing screws is a great choice because they come with a washer, are strong and compatible with impact drivers (which tend to break regular Phillips head drywall screws), and they

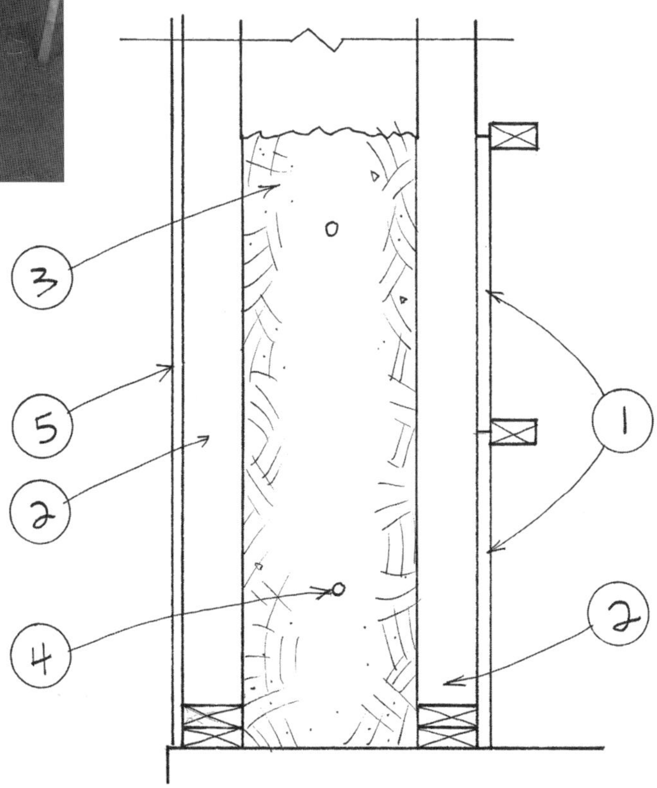

1. Strongback form
2. Stud
3. LSC
4. Horizontal reinforcement rod
5. Sheathing as form

Formwork Fasteners

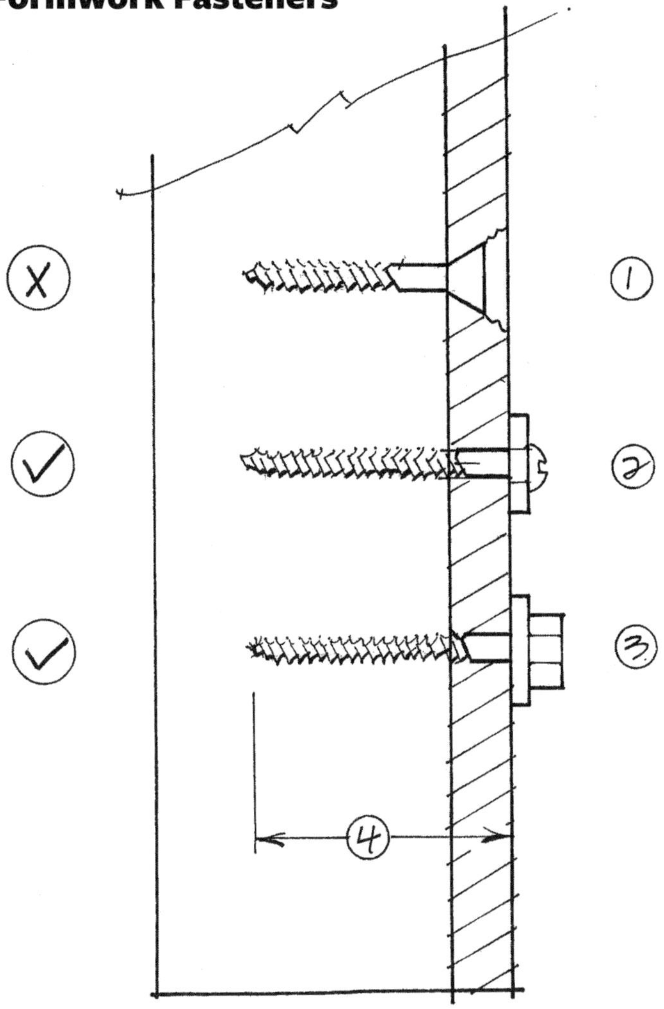

1. Tapered head screw. These will inevitably be driven too deeply into the forms and will be difficult to find and remove, especially once the forms have some slip on them.
2. Pan head screw with Robertson or Torx head. These will stay on the surface of the form and be easy to locate. The heads may fill up with slip over time.
3. Hexagonal head screw. These will stay on the surface and will not be affected by slip.
4. Choose a screw length that is 2 times the thickness of the form material (½ inch [12.5 mm] forms use 1¼ inch [32 mm] screws). Screws that are too long will waste time every time they are driven and removed.

can be reused with ease. The other advantage to the hex head fastener is that it cannot get full of clay slip, like a standard female screw head can. Torx head screws and timber lock screws are also durable screws for a straw clay project.

A typical strongback form is 16–24″ tall by 8′ long. This is the most efficient use of plywood. Forms can be cut to specific heights and wall lengths. For a typical residence where the game plan is to infill the whole structure at once, you will need to have two forms for the interior side and two forms for the exterior. This generally works out to linear feet of wall divided by 8′ for the number of sheets of plywood you will need. In some cases, a full sheet is placed on the non-installation side of the building and the forms are leapfrogged only on the interior or exterior of the building to be more efficient with the time it takes to leapfrog the forms.

Dimensional lumber can also be used as formwork, however it is much more efficient to fill a space that is 12–24″ deep, which is hard to achieve quickly with narrower dimensional lumber.

Forms can be removed immediately after installing the straw clay. For the first batch of straw clay, immediately remove the form to inspect the work and assess if more tamping or less is needed. It is best to not leave forms attached for more than 48 hours; leaving them on any longer may encourage mold. When leapfrogging the forms, it is best to fill all the way to the top of the top form before moving the bottom form up. It is also possible to only use one form and move it incrementally up the wall as it is filled. To get a solid wall, it is important that you not place the form exactly level with the last level of installed straw clay. Keep the form overlapping the previous layer by a minimum of 4″ to avoid the wall bulging at the seam.

An important tip with temporary formwork is that when tamping the straw clay in, focus on the edges and corners. This locks the straw clay into the face of the cavity and prevents pockets from forming that will have to be filled in later.

For any project, one person should be designated as the "clean hands person" who operates the drill and moves the forms. This will keep things running smoothly and will add to the longevity of your drills. Remember: clay slip (i.e., water and soil) is the enemy of tools with moving parts. If you are working by yourself or in a small group, keep a bucket of water and a towel available for people to clean and dry their hands prior to using the drill.

Formwork for curving walls

It is possible to make temporary formwork for curving walls. For a gentle radius, ½″ plywood can be bent in place and screwed to the structure (there is generally an increase in vertical wood members in a curving wall section, so there will be lots of points to fasten to). For tighter curves, a kerf cut every inch in the plywood could suffice; or MDF (medium density fiberboard) or some other flexible material could be used. If you have a large circular building, it might be worth the effort to make curved strongbacks for the plywood. Or it might be worth it to consider using a permanent formwork.

There are a variety of ways to handle curves around openings like windows and doors, most of which implement permanent formwork. See Chapter 7 for more details on openings. For temporary formwork, using Sonotubes can be an affordable and efficient way to handle this detail. Sonotubes are cardboard forms used for concrete to pour piers and circular pads; they come in a variety of diameters. Because screws tend to blow through them, if the plan is to reuse

Formwork for Curving Walls

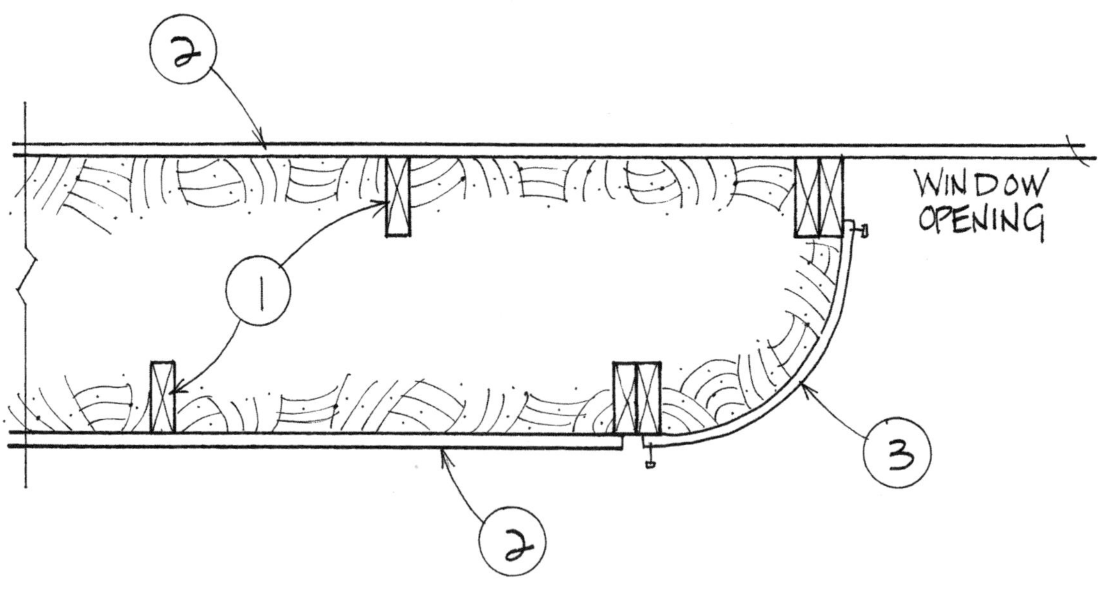

1. Studs
2. Strongback form
3. Sonotube

the Sonotubes, fasten the edges with a strip of wood lath or plywood for longevity. For 12″ walls, a 12″-diameter Sonotube can be cut in half or quartered to achieve the same radius around each opening. They can be installed prior to filling with LSC and removed after the filling is done on both vertical and horizontal curves around window and door openings. They can also be used to achieve curves both on inside and outside corners, creating both convex and concave curves. ABS or other plastic piping can be cut to make forms as well.

Permanent Formwork

There are a variety of types of permanent formwork for light straw clay buildings: wood lath, wire lath/mesh, reed mat, and split bamboo mat are some of the choices.

Permanent formwork stays on the framing and becomes part of the plaster substrate. Advantages of permanent formwork are that it goes on and stays on, so you are not "wasting" plywood. It allows for a much lighter mix to be infilled in the wall cavities, creating shorter dry times and requiring less material. Depending on the situation (climate, wall thickness, density of mix), it can often be plastered over immediately.

Wood lath is a traditional way of creating a substrate for plaster using small strips of wood that are usually ½″ thick, no wider than 1¾″, and spaced no more than ⅝″ apart. See Chapter 8 for installation details.

Wire lath/mesh formwork is used much like wood lath except that it is rolled out, and, depending on the wire and structural member spacing, it is sometimes installed in two layers. There are a variety of metal meshes on the market: expanded metal mesh (also affectionately referred to as "blood lath," due to its tendency to lacerate the installer) in a variety of weights/strengths; chicken wire; fencing wire; and welded wire mesh. It is possible to design for your welded wire mesh permanent formwork to also provide the shear strength in your structure, providing for two functions at once.

Wire mesh embedded in lime plaster can also be designed to provide the shear strength in a wall system. This has been used in straw bale construction and can be used to create shear walls for straw clay. Lime plaster has wonderful qualities: it is harder and has a higher compressive strength than earthen plasters; it is water resistant; and it is relatively easy to apply and acquire. See Chapter 8 for installation details.

Reed mat is a roll of small reeds that are held together with a thin wire. These rolls generally come in 6′ × 15′ rolls and are stapled with the reed running perpendicular to the structural

Reed mat in place. PHOTO CREDIT: LYDIA DOLEMAN

members. Rolls can usually be purchased in garden supply centers. Due to the fragile nature of the reeds, it is recommended to use a roll of wire or baling twine to staple over the reed mat when affixing it to the studs so the reeds don't break. The reed mat is very flexible, so it is not recommended for use when cavity spaces are over 24″ wide. It is cut with a utility knife or scissors and provides great key for plasters to be applied over. It will bulge if the light straw clay is overtamped, and special care has to be taken with detailing around electrical outlets. See Chapter 8 for more details.

Split bamboo mat is much like the reed mat, but is made of ridged sections of bamboo that have been split. It is also sold in 6′ × 15′ rolls and installs exactly like the reed mat. The advantage to the more expensive bamboo mat is that it is much more rigid, can span a bigger area, and has less of a tendency to bulge. It is harder to cut, though; it requires shears, a hand saw, chop saw, or a circular saw.

Chapter 7

Design Notes, Details, and Budgeting

DESIGN IS A FIELD ALL ITS OWN. Good design is like calculus: there are lots of variables and lots of different potential outcomes. For most projects, the ideal guiding principle is ethical, ecological, and sound design, but in reality, it is most often budget and cost that ultimately control the reins.

Designing for Plasters

Given that most LSC buildings are clad in plaster on the interior, the exterior, or on both sides, it is important to factor this in during the design phase.

Questions to ask during the design phase:

What kind of treatment will there be around interior and exterior windows? Wood trim? What kind of sills, inside and out? Will there be base trim? How will plaster meet exposed framing in a timber frame or pole frame? Will J-channel be used? How thick is the planned total plaster to be on the interior and exterior? What type of outlet covers will be used? What type of lights will be mounted? In which rooms will the walls need to be washable? What type of look is desired for the plaster: smooth and reflective? Rough? Skip-troweled? Should there be zero evidence of trowel marks, or subtle trowel marks? Will walls be in high traffic areas and need constant repair work? What will be mounted to the walls (cabinets, shelving, artwork), and where? Will there be a need for picture railing to hang decorations? Is there a desire for special features, like truth windows, niches, or bottle walls? Is there a desire for rounded corners on interior and exterior corners? What level of skilled labor will the budget allow for? What colors might be desired?

A design calculus equation:

A well designed home = time to research and experience different spaces + good design for the specific location (i.e. climate and inhabitants' lifestyles) + local code requirements + appropriate material choices + (realistic desires of inhabitants × capacity for decision making/by amount of time available) + desired aesthetics + building season timeframe — all divided by budget.

Answers to all of the above questions will inform plaster details and prep.

A Note on Air Fins

Where plaster meets non-plaster materials such as wood, windows, doors, and pipes, gaps will open up when the plaster dries. Installing *air fins* at these junctures — materials such as tar paper, fiber board, building wrap, and specialty sealing tapes — is essential to a tight building envelope. Prepping for a very tight thermal envelope should be done prior to plaster prep and certainly before any coats of plaster are applied. Placement of air fins should be delineated on the plans and the use of tapes or cut tar paper or stucco paper (ideally, two layers — especially if using a lime plaster) should be mapped out in advance. Air fins should be placed at the top of the wall prior to hanging a sheet rock ceiling (highly recommended to be installed prior to plaster). It is best to install your air-fin tape or paper in the same manner as you would roofing underlayment: from the bottom up. Have each layer shingle over the next. This is good practice and a good preventative in the event you have some sort of water issue in the future.

Details for Plaster Skins at Top and Bottom of Walls

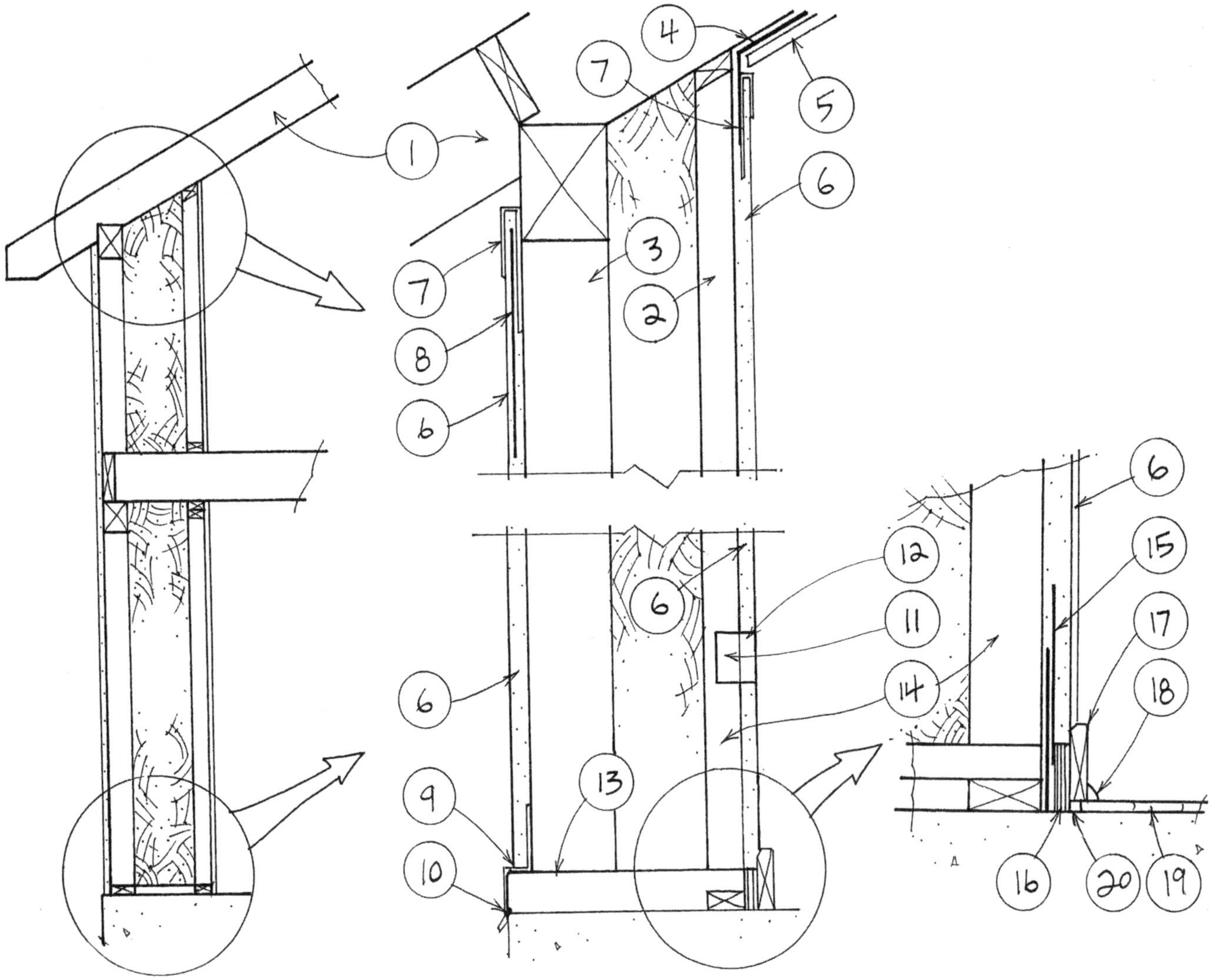

1. Ceiling joists
2. Secondary frame
3. Superstructure
4. Air fin
5. Sheet rock
6. Coats of plaster
7. J-channel
8. Building paper
9. Flashing
10. Bead of caulk
11. Electrical box
12. Box extension
13. Double bottom plate
14. Stud
15. Lapped building paper
16. Plywood spacer
17. Base trim
18. ¼″ round over trim
19. Flooring
20. Gap

Top: *LSC 200 square foot pole frame building with earthen plasters. Note the living roof on the south-facing roof.* Photo credit: Erica Ann Bush

Above: *Same LSC building. Note the sculpted plaster around windows in the center and the two different foundations styles: stone staked and mortared; and lime plaster over concrete stem wall.* Photo credit: Erica Ann Bush

LSC wall system with curved radius earth-plastered window openings.
PHOTO CREDIT: ERICA ANN BUSH

Top: *LSC residence in southern Oregon. Post and beam frame with round wood porch posts and accents. Earthen plastered. Pump house is also LSC and earth plastered.*
PHOTO CREDIT: ERICA ANN BUSH

Left: *Timber framed EcoNest LSC building with earthen plasters and horizontal wainscoting.*
PHOTO CREDIT: ERICA ANN BUSH

Left: *Portland, Oregon, permitted LSC retrofit of an existing residence.* PHOTO CREDIT: MARK LAKEMAN

Bottom: *Interior detail of the LSC retrofit. Note the open sills and the application of LSC on the interior and the exterior (previous photo) to maintain shear in the building and facilitate drying to one side and then the other.* PHOTO CREDIT: MARK LAKEMAN

A window added to a residence where plaster was cut into and a new window fills the cavity. Note the deep sill and minor repair work to the plaster.

Photo Credit: Lydia Doleman

Combination of pole framing and conventional 2x8 framing. Exterior is clad in locally sourced wood siding coupled with a rainscreen.

Photo Credit: Lydia Doleman

Left: *Sprouts! A common occurrence and a great conversation starter about LSC walls. Seed heads in the straw sprouts (though not usually with rice straw). As the sprouts grow, they help pull moisture out of the wall . When they wilt and die off, it is an indication that the walls are drying well.*
PHOTO CREDIT: ERICA ANN BUSH

Below: *Larsen truss building under construction. Designed to be infilled prior to installing roof. Note the Larsen trusses on the south wall and full studs on adjacent walls.* PHOTO CREDIT: ERICA ANN BUSH

Bottom: *120 square foot timber frame LSC auxiliary bedroom. Timber frame superstructure infilled with Larsen trusses. All reclaimed wood. Lime plasters on the exterior and shingled lower sections to protect from Portland, Oregon's consistent rains.*
PHOTO CREDIT: LYDIA DOLEMAN

There is a wide variety of colors available from iron oxide pigments. Photo credit: Lydia Doleman

Mixing slip can be a luxurious process. Note the swirls of color, indicating lots of clay particles in suspension. The dimples on the surface indicate that more water could be added. Photo credit: Lydia Doleman

Radius trowels help keep an even radius when doing interior and exterior corners. Photo credit: Lydia Doleman

Details: Window and Door Openings and Electrical and Plumbing

Window and Door Openings

There are a lot of ways to detail window and door openings. In most instances (unless in a consistently very hot, sunny climate) windows and doors are set to the exterior of the building, simplifying the sill details and creating deep window wells on the interior of the building. This also helps keep moisture and rain on the outside plane of a building. A large sill in the center of a wall has the potential to leak into the middle of a wall, which can create hard-to-find-in-time problems. Also, detailing to prevent water infiltration requires a lot of foresight and a high level of skill. Given the wide variety of window details, for simplicity's sake we will stick with detailing the light straw clay.

To achieve curved window openings, there are a variety of options. The most basic is to make a very high-slip-content light straw clay mix and sculpt it around the window openings. Nails or other mechanical means are used to bond the straw clay to the vertical wood members. It is important to decide prior to starting infill of the walls what kind of windows you will be installing and what kind of trim they will have, as that will determine where the straw clay should dive into the wall by the window.

Another method is to use a Sonotube or piping of some sort (PVC, ABS, etc.) that is cut and fastened to the window to create a uniform radius form. This is especially useful in making the top of a curved window opening easier to deal with (the tendency is for gravity to take over, causing the straw clay to sag or just plain fall off). Using wood lath is a great way to achieve a flawless and uniform upper curve. This technique can also be applied to the sides of the windows and around doors.

For an angled opening, the same techniques as above can be used but instead of pipe or Sonotube, a form of plywood can be made and removed after filling. On occasion, this technique is also used for exterior and interior corners depending on the framing.

Use wood lath to create a solid, symmetrical upper radius above windows and doors. It is also a great use of leftover scrap wood.
PHOTO CREDIT: LYDIA DOLEMAN

An affordable and easy way to detail a window.
PHOTO CREDIT: JIM RIELAND

Cross-Section Details of Different Window Openings

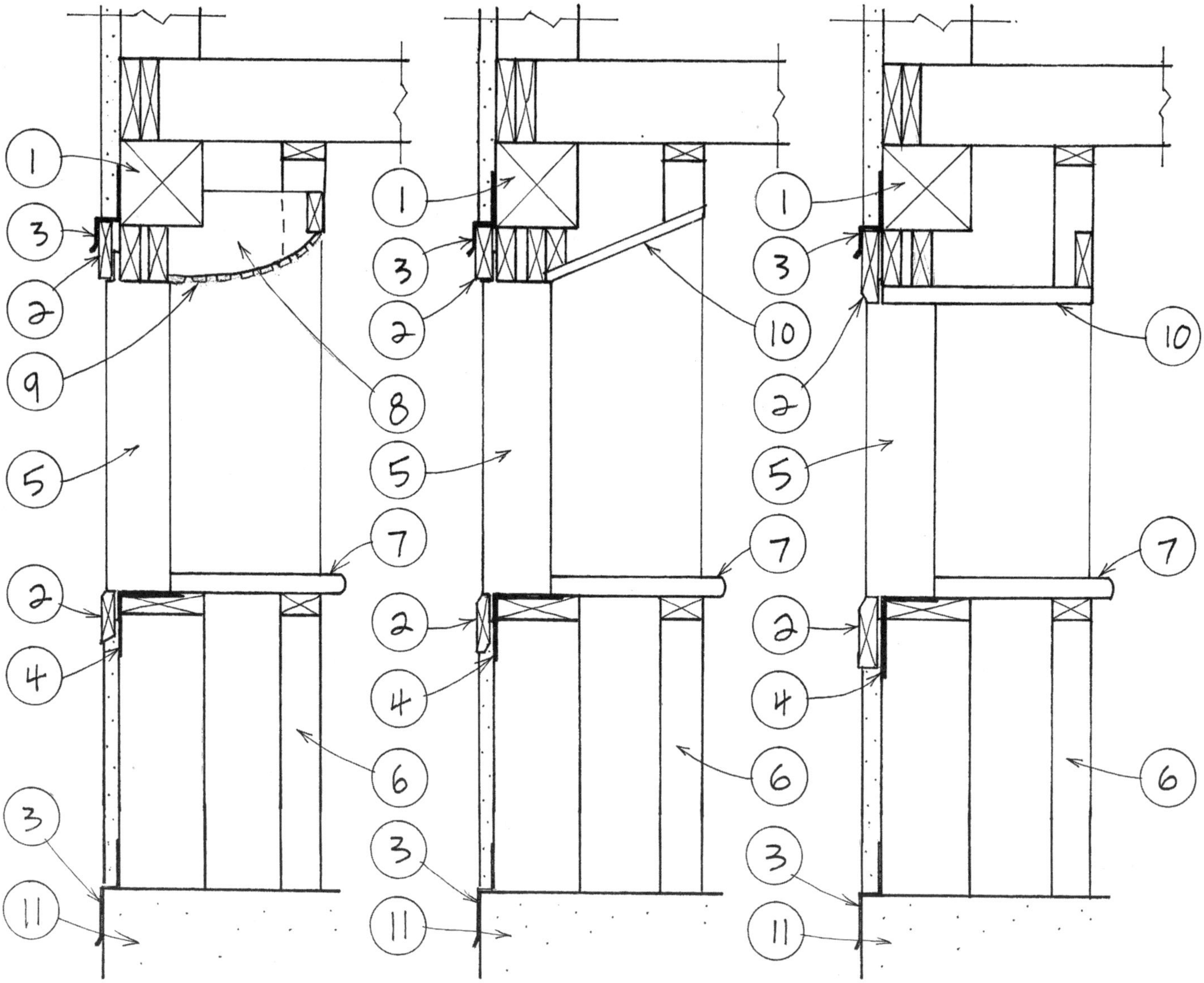

1. Framing
2. Trim
3. Flashing
4. Sill seal or ice and water shield
5. Window with flange
6. Secondary interior framing
7. Sill
8. Rib to support lath
9. Wood lath
10. Plywood box
11. Foundation

Plan View of Rounded vs Boxed-in Window/Door Openings

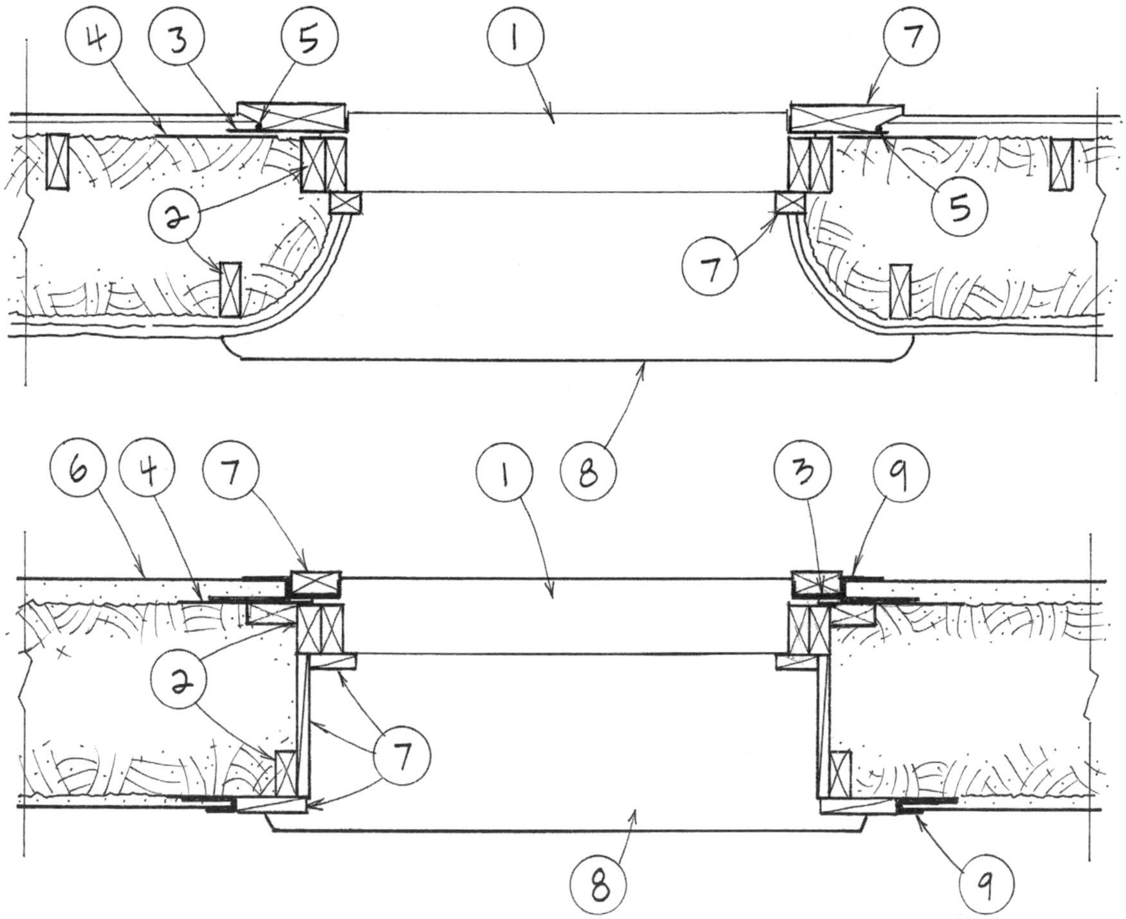

1. Window
2. Framing
3. Ice and water shield
4. Building paper
5. Bead of caulk
6. Coat plaster
7. Trim
8. Sill
9. Channel

Radius trowels help keep an even radius when doing interior and exterior corners. PHOTO CREDIT: LYDIA DOLEMAN

If a plastered window or door opening is not desired, it is very quick and affordable to make a window or door opening entirely out of wood. This can be plywood that is later trimmed out or whatever wood that works with the chosen aesthetic of the project.

A Note on Openings

Sharp corners tend to break. Curved corners tend to resist chipping much better. Interior sharp corners can make good stopping points and are easier to keep uniform unless you are using a radius trowel.

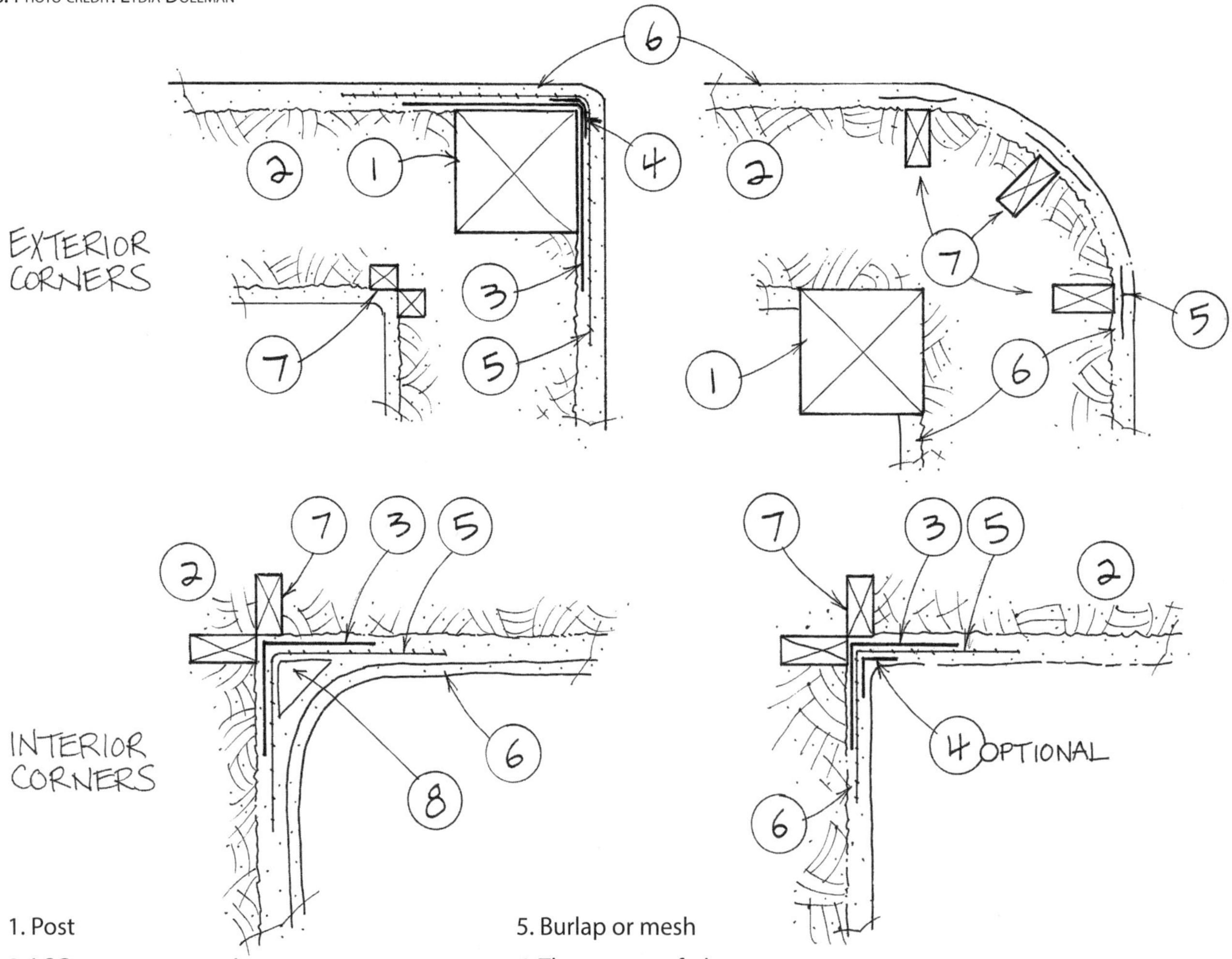

1. Post
2. LSC
3. Building paper
4. Corner reinforcement used for sheet rock
5. Burlap or mesh
6. Three coats of plaster
7. Secondary framing
8. Plaster buildup for radius

Top of the Wall and Odd Spots

As described in Chapter 8, at the top of a wall, a mini form is needed and tamping needs to be done from the side. The straw clay mix might benefit from a higher slip content in order to make installation easier. It is also advisable to install a "key" in the framing portion along the top of the wall. Keys at top of the walls and in stud bays mitigate shrinkage and deter drafts.

For the top of gable ends, the same trick applies — though it involves tilting forms at an angle and tamping from the low end to the high end.

For gable ends or angled walls and permanent formwork, it may be simpler to pick an ending point and run reed mat parallel to the ceiling for ease of installing the reed mat.

Reed mat can be applied in oddly shaped spaces.

Photo credit: Lydia Doleman

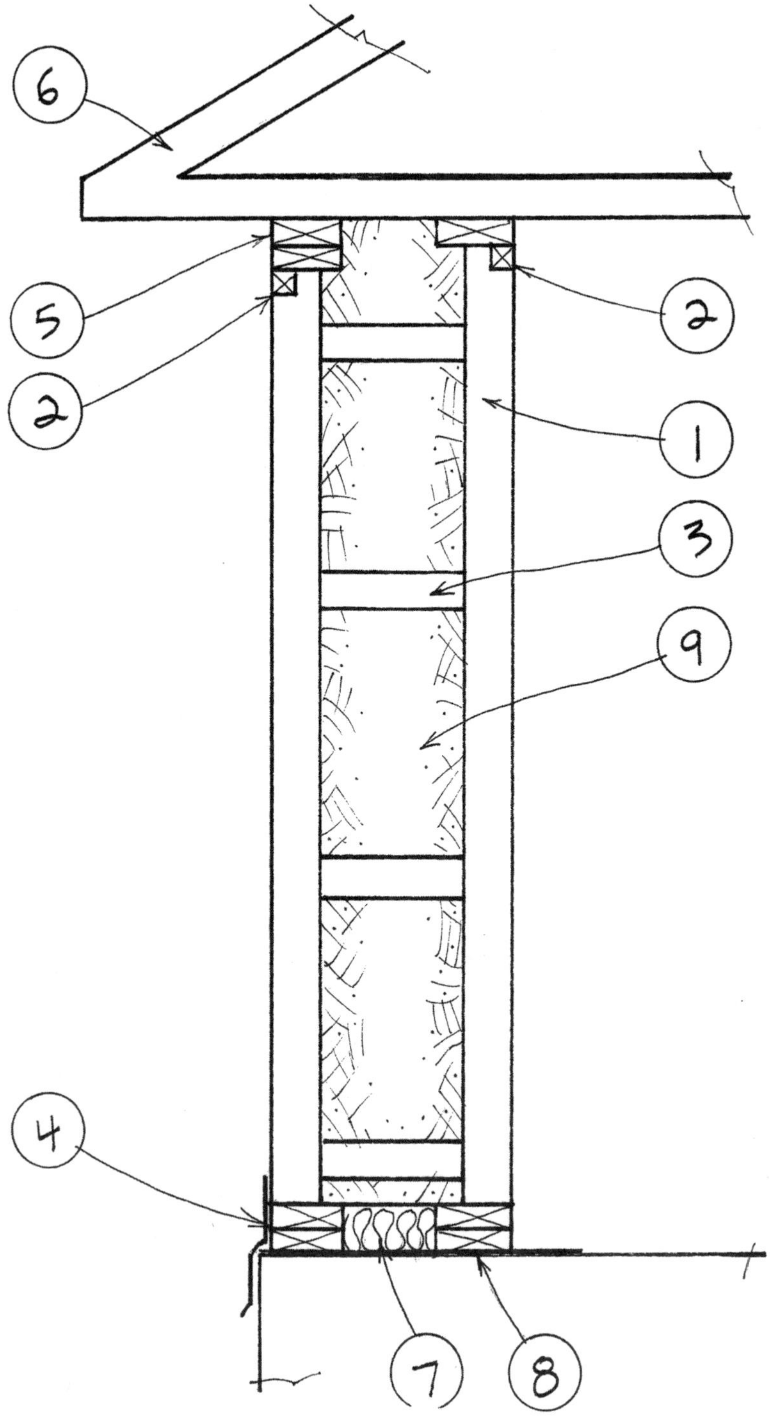

1. Larsen truss
2. Key
3. Gussets
4. Bottom plates
5. Top plates
6. Roof truss
7. Rigid insulation
8. Vapor barrier
9. LSC

Electrical Details

For temporary formwork, hold boxes flush to the exterior of the interior framing; instead of cutting out for every box, switch, and light, having all boxes set flush makes the light straw clay installation go much faster. Jam extensions (plastic extensions of the electrical boxes that are sold in a variety of depths and sizes) can be put on later.

Electrical boxes are mounted flush to framing so formwork can fit seamlessly. Holes in boxes are taped to keep LSC out. Photo credit: Lydia Doleman.

For rooms or projects where expansion or remodeling is anticipated, it is wise to run all electrical in conduit. It is an added initial expense, but it's significantly cheaper than trying to channel out for new wire later. If you install conduit, it will be much easier to later add a circuit or rework any of the electrical.

Where electrical runs through the studs or members, it is very important to install plate covers; this protects the electrical should long screws be used for anything (think how many screws are used in the forms alone). It is also a good idea to photo document the electrical in the wall cavities prior to infill; print the photos and keep them for your records. Don't rely on your phone for photo storage. Your walls will surely outlast your current phone, and photos are easily lost in the transfer process to a new phone.

For temporary forms, mount electrical boxes to studs and create a wooden surround to staple lath or mesh or mat to. This surround can be let-in plywood or dimensional lumber. The box can be set to the anticipated finish plaster depth (though setting a box just shy tends to result in better final plasters as a box a little proud can leave shadow lines and catch dust on an otherwise flat wall).

Be mindful of electrical wires while tamping and nailing. Nicking a wire could be hazardous. Keep wires as close to the middle of the wall as possible.

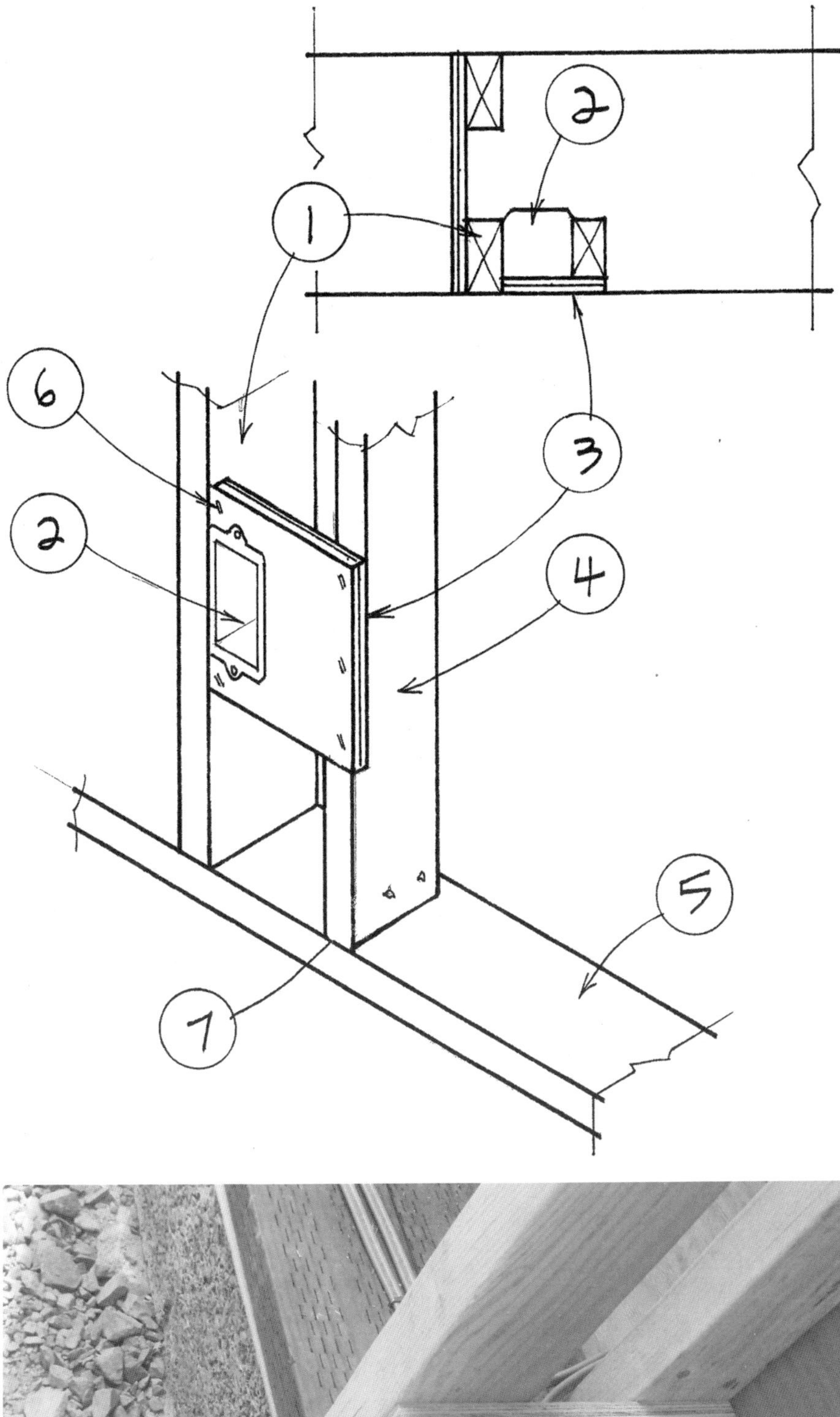

1. Stud/vertical member
2. Electrical box
3. Plywood
4. 2×4 notched support
5. Bottom plate
6. Staples
7. Nails

Gussets of Larsen truss are held up so wiring can run underneath, alleviating the need to drill holes.

Photo credit: Lydia Doleman

Plumbing: Water and Gas Lines

If possible, try to design the plumbing and vents into interior walls and keep the water out of the envelope wall. If this isn't achievable, be sure to document where the plumbing runs and double sleeve it so that a leak won't go into the wall, but will be detectible elsewhere. Double sleeving is generally recommended for water lines and drains, but not vents, as they are not there to drain water. All water plumbing should be run prior to light straw clay infilling. Some jurisdictions will not allow plumbing to be run through exterior walls. Even if it is allowed, though, if you live in an exceptionally cold climate and plumbing in the exterior walls is not prohibited, consider moving plumbing to the interior for better freeze protection.

Gas can also be double sleeved in a wall, but is less vital than double sleeving a water line. Given that gas lines are usually stubbed up through the foundation or run directly through an exterior wall, it is important to both have a form ready to be notched around the pipe and also to be sure that the pipe on the exterior is under an overhang so that it is protected from rain; otherwise, it can become a conduit for water to get into the wall system.

As for your electrical, photo document where the plumbing runs in a wall; in the event of a problem or a renovation, you'll want to know where the pipe runs.

Budgeting and Materials Calculations

Each light straw clay building is unique — on many levels — making it very challenging to estimate an average standard cost per square foot. In North America, the costs for average-size (1,500–2,500 square feet) conventional homes (not owner/builder) run about $125/square foot to $250/square foot. This is a ballpark starting point for mapping out a budget for a single family residence. The more design features added, of course, the more the cost will go up. It is not uncommon for a custom natural home to be $250/square foot or more. If you are about to faint, and your dreams of building an affordable home have been dashed, fret not. There are a lot of achievable ways to rein in the cost of a newly constructed naturally built home.

One of the best ways to cut costs is to take your time and design for simplicity. Complicated foundations and roofs add big costs to a building. A simple rectangle that is one or two stories tall saves in materials and engineering. You can also make a modular design with the plan to add on in the future.

Home construction costs can be divided into materials and labor. The average house is about 50% material costs and 50% labor costs. You can build it all yourself and save 50% of the costs if you have the time and skills to do so. Or, you can take the time to find materials at cost or below cost; if you stockpile them, you can then design around them to help save money. However, a crucial element to consider with a natural building (or with any residential structure) is that there are upfront (construction) costs, and there are lifetime costs (heating, cooling, and maintaining). These lifetime costs can far exceed the construction costs if a building is not designed properly.

Wall systems generally only represent about 10–15% of the total building cost. A light straw clay building still has a standard roof and foundation system that any residential building would include.

Budgeting for Materials

Straw bale quantity

2-string = 16″ h × 18″ w × 40″ l = 6.6 ft^3 of compressed straw

3-string = 18″ h × 24″w × 45″ l= 11.26 ft^3 of compressed straw

round = 5′ × 5′ = 98 ft^3 of compressed straw

Bale dimensions vary. Check with your bale supplier, and make adjustments to calculate the cubic feet in the baled straw you'll be using.

To calculate how much straw to get for a 12″ wall system, first you must tally up all the wall volume to be filled with straw clay:

- Length of wall × height – window and door openings = square feet of wall space to be filled.
- Take your square feet of wall space and multiply by 0.5 if a 6″-thick wall, or 0.75 if an 8″-thick wall, or by 1 for a 12″ thick wall.
- Add 10% just as a buffer to insure more than enough material; it is never worth it to estimate low and end up running shy of materials during the installation. Generally, it is best to do the calculations to the point of tabulating the exact number of bales, and then round up 10%.

For example:

You have 1,200 square feet of wall space (200 linear feet of wall × 10 foot wall height minus 800 square feet of window and door openings = 1,200 square feet of wall space). This would equal 1,200 cubic feet of wall space to infill for a 12″ thick wall system. For an 8″-thick wall system, the 1,200 square feet would be multiplied by 0.75, to get 900 cubic feet of wall cavity.

Each 2-string straw bale on average contains 6.6 cubic feet of compressed straw. When unfurled from its baling twine and loosened to go into a tumbler or tossed on a table, straw expands to nearly three times its original volume. Given that it will be semi-compressed back into the wall system, an easy estimate is that the cubic volume in a stringed straw bale will be twice that volume in the wall.

Thus, one 2-string bale gives approximately 13.2 cubic feet of wall infill.

For our example wall system that is 12″ thick, you would need 100 bales:

1,200 square feet = 1,200 cubic feet for a 12″-thick wall divided by 13.2 = 90.09 bales + 10% = 99 bales — or 100 for an easy round number.

For an 8″-thick wall, you would calculate:

1,200 square feet = 900 cubic feet for an 8″-thick wall divided by 13.2 = 68.2 bales +10% = 74 bales — or 75 for an easy order number.

For the 8″ thick wall, where you've already rounded up to 75 bales, a *squeeze load* of 100 bales might still be the most cost effective. Twenty-five bales could be used for other purposes like scaffolding or seating, or could be resold if kept in good condition. Depending on your location, extra bales can be an asset. However, if you have a very tight building site (i.e. urban area), the challenge of storing extra bales may outweigh the benefits of having them on site.

Calculating straw costs

Number of straw bales × cost per bale + delivery fees and/or squeeze rental = total cost of straw bales

Calculating clay requirements

Prior to commencing the LSC wall infill, it is necessary to map out what quantities of clay are needed to achieve the wall density you are looking for.

All calculations are based on achieving a 13 lbs/ft^3 (pcf) density of dried LSC (6.7 lbs of straw and 6.3 lbs of clay) for a thermal resistance of R-1.69/inch.

To reiterate, in addition to the amount of clay, density is ultimately controlled by the amount of tamping. Before starting in on infill, you should make test blocks (see Chapter 3, "Creating Test Blocks to Assess LSC Mix Density" and Chapter 8, "How do you know if you tamped it right?") and ensure that everyone involved in the tamping process has practiced and has a solid grasp on the degree of tamping that will achieve the right density.

Calculating Amounts of Bagged Clay

One 50-pound bag of ball, fire, or brick clay will make sufficient slip to cover 7.9 cubic feet of LSC (50 lbs/6.3 lbs per ft^3 of clay in the LSC mix at the optimal density of 13 lbs per ft^3 of dry LSC).

1 bag = 7.9 ft^3 of LSC.

For 1,200 ft^3 of infill:

1200 ft^3/7.9 = 152 bags of clay

To ensure sufficient materials on site, 10% should be added.

Calculating bagged clay costs

Cubic feet of wall space/7.9 = minimum number of 50-pound bags of clay + 10% × cost per bag + delivery fees.

Calculating sifted site subsoil requirements

Once again, base your calculations on your own testing. The following is based on the author's averages and data from other builders. Given that site soil doesn't come in 50-pound bags, and is usually delivered or quantified in yards, the following calculations are based on cubic yards of material needed. Also, please note that site soil ranges in densities, composition, and clay contents. Variables include how finely sifted it is, moisture content, amount of silt and aggregates present, and types of clay. Taking the time to create test blocks is required to insure a proper LSC wall system (see Chapters 3 and 8 for more on test blocks).

One 5-gallon bucket of sifted soil = 0.66 ft^3 of sifted subsoil.

One 5-gallon bucket of sifted subsoil can equal 8 gallons of slip.

Thus: 0.66 ft^3 of soil = 5 gallons of soil = 8 gallons of slip. This is the amount of slip needed to coat 6.3 ft^3 of straw to a dried LSC wall density of 13 lbs per ft^3 (6.3 lbs soil and 6.7 lbs straw).

In larger volumes, a 50-gallon barrel of slip would thus contain 4.125 ft^3 of subsoil (6.25 5-gallon buckets of subsoil) and will coat 39.6 ft^3 of straw, which is three 2-string bales. (Note: a 55-gallon barrel is rarely ever filled to the top, making it more like 50 gallons of slip.)

To work the example to determine the number of 5-gallon buckets of subsoil required and thus the cubic feet and cubic yards:

1,200 ft^3 of wall/6.3 ft^3 of straw = 190.5 5-gallon buckets of soil × 0.66 ft^3 per bucket = 125.4 ft^3 of soil/27 = 4.64 cubic yards of sifted subsoil.

To ensure sufficient material on site for the duration of a build, the volume of pre-sifted subsoil should be increased by 15–20%, which will also compensate for any loss of material during sifting.

Calculating site-sifted subsoil costs

Cubic yards × cost of subsoil per cubic yard + delivery fees + labor for screening (or labor/machine costs for digging and screening).

Chapter 8

Construction Procedures

ONCE A BUILDING IS FRAMED and the roof is on, or "dried in" (meaning the building is essentially dry under the roof), or if you are opting to leave the roofing portion until after the walls are filled, it is time to get your infill materials prepped. Sometimes this can happen prior to and sometimes during the framing process.

Straw

If you are getting over 100 bales, delivery is great, and hiring a squeeze operator (a cool piece of drivable farm equipment that literally "squeezes" a block of 100 bales and can drive where a semi truck can't) is even better. These wonderful machines can save you hours of labor and many backaches over the long haul. If those aren't available, plan to rent or borrow a large truck or flatbed, bring some good gloves, and, unless you are up for a workout, some help. Stack bales for travel in an interlocking way so that they are both stable in transit as well as while you are unloading them. Straw bales are great, and an ingenious way to move a lot of straw in a compact way, but appreciation for a bale wanes after the third time moving it (especially those heavyweight champion 3-string rice straw bales). Plan well, stack well, and, well, save your back for the rest of the house.

Another thing to note about acquiring bales is that for a light straw clay project (unlike a straw bale project) bales that have been in a barn for a year and stored dry are perfectly acceptable to use and sometimes more affordable because some farmers really need to make space for the next crop to come in. The downside of older bales is that since they provide such fabulous insulation and the occasional seed head, they make ideal habitat for field mice and other critters that you wouldn't want in your wall; you might be shocked when you open one to find a family of 12 mice in there before you mix it in the clay slip!

Light straw clay can be installed without a roof on for quicker drying — if you have predictable weather during your building season.
PHOTO CREDIT: ERICA ANN BUSH

A bale that has been inhabited is still usable as long as the vast majority of it seems dry (still golden, feels dry to the touch). The only downside might be the potential of *hanta virus,* which is spread via mouse urine. Although it seems to be extremely rare, it is worth a mention — particularly if you are in an area prone to that virus. Using bales is a dusty affair; so wear a dust mask, regardless.

If a bale has a dark green spot or is significantly heavier than all the rest, it is likely higher in moisture content and is best left out of the mix; alternatively, you could just discard the discolored section after you cut the twine. Anything that has a moldy smell or shows evidence of previous water damage is likely to inoculate the mix with more mold spores than are already there and might contribute to mold problems if conditions are right (or wrong, depending on how you look at it) for mold growth. Scrutinize your bales! You can be comforted by the knowledge that any water-damaged "unusable" bales have a variety of secondary purposes: scaffolding on site, mulch, or seating.

Storing Bales

If you are lucky enough to have the bales delivered the day before you plan to install the straw clay, by all means just plop them near your mixing station and have a tarp big enough on hand to cover them up in the event of a rainstorm. However, most folks get their straw when it is available and need to store it for a week or more prior to using it. Storing the bales in a barn or shed for the duration, or even inside the project itself is a possibility; however they take up a significant amount of floor space, and you might need to move them around as the project progresses. If at all possible, put them somewhere where you won't have to expend labor to move them repeatedly.

If you need to store bales for any amount of time and don't want to have to move them again, place pallets down or a grid of 2× material on the ground to keep the bales off the ground in a spot near the intended mixing site. The moisture in the ground will migrate up and saturate the bottom course of bales if you don't provide an airspace; even then, the bottom course of bales will feel significantly heavier despite being off the ground. Even dropping a tarp will make a difference (just like camping), but airspace is the best if you can't do both. Stack bales as densely as possible and interlock them, so the pile is secure. If you can, create a spine of bales, or of some wood to create a ridge for your tarp materials to drain on either side — instead of pooling water on top. Then tie the tarp down securely (think windy days); the stack can sit for as long as your tarp stays intact (if storing for more than a month, double-tarp the pile, as the UV will break the tarp down quickly). Plastic sheeting, EPDM (pond liner), or any non-vapor/non-water permeable membrane is usable.

Straw bales vary in cost, as a large portion of the cost is transportation. In southern Oregon in 2016, prices for locally grown wheat straw averaged $5.75 to $7.75 per 2-string bale (delivered). Three-string bales averaged $7.75 to $9.75 per bale. Ordering quantity can make a difference as it is much more efficient to fully load a semi than only partially load it. If you are hiring a squeeze loader (sometimes this is factored into delivery costs and is worth inquiring about if ordering more than 100 bales), this is also part of the cost of delivery and transport.

Processing Site Subsoil and Clay

As straw comes nearly ready for mixing, the biggest time-consuming portion of an LSC building is prepping the clay — if you aren't purchasing bagged clay. Depending on where you are ultimately sourcing your clay from, it will be worthwhile to map out how much time and money will be put into processing imported or site-excavated clay subsoil compared to the relative simplicity of using bagged clay.

In most cases, when the clay arrives, or is excavated prior to infill, it needs to be kept dry. There is nothing harder than taking big, soggy clumps of clay and trying to get them to the consistency of a milkshake for the clay slip — particularly if your subsoil has a high clay content. In the event that your pile is soaked, or you have no choice but to take a wet load of clay, you have two choices: let it dry out and then sift or pulverize it; or soak it in containers and mix it. I recommend the latter. Drying it out and trying to break apart clumps of dried clay is hard work. If you do have the time to dry it out, that might be the best option, but if you have a short window between when the clay got wet and when you plan on infilling your LSC walls, you will have to work with it wet. Unless you have access to an industrialized sifter (you can rent one, and it might be worth it), you will be spending a lot of time drying and sifting something you are only going to get wet again.

However, if this is the route you must go (for instance, if your soil has clay and lots of rocks, and was dry but got rained on), then wait for a sunny or at least dry day and start sifting. If the pile had gotten wet and then dried, and it is now a giant, cohesive, dry lump, mist it with water until it softens, and you can slowly get the soil to a workable level of moistness without it being big, gooey clumps. If you inadvertently end up with clumps, follow the instructions for soaking given below.

A tip on sifting: just go for the easy stuff. If you have a mountain of a pile, do a quick sift first: just throw the soil against a screen (½″ mesh is good enough to remove enough large rock and debris for subsequent pumping of slip if a tumbler is being implemented) and maybe scoop up what won't go thru two to four times then toss to another pile. You can sift that more carefully later if you need it. Or, you can spend the time the first go around trying to squeeze as much soil out of the first round as possible. I usually decide on which approach to take depending on what the labor pool is like for the

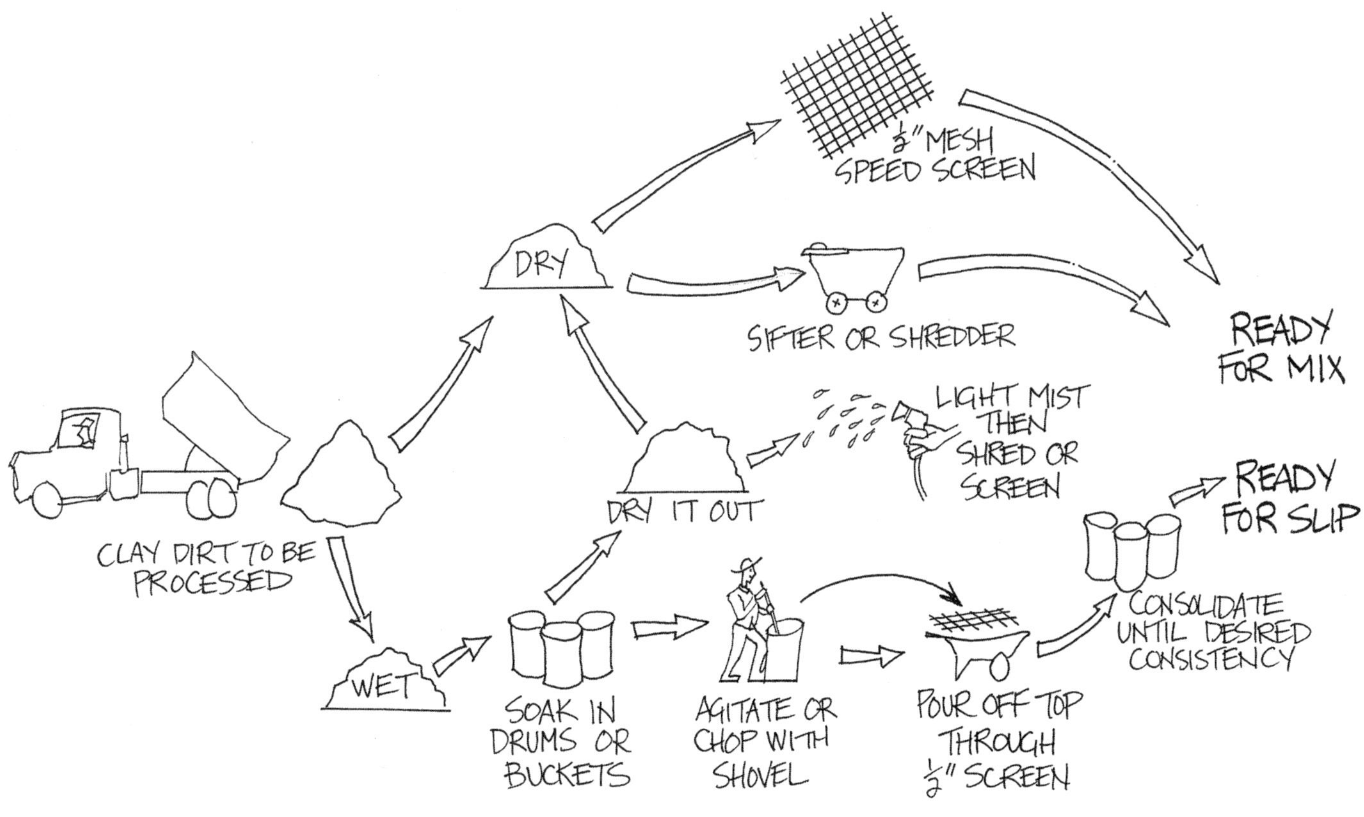

CLAY PROCESSING

day. If there are a lot of people who want to sift, then sift by golly! If it's just you on the business end of a shovel, then aim for the proverbial low-hanging fruit and save the hard stuff for another (maybe not necessary) day.

To soak: take big chunks and place in the bottom of buckets or barrels or in a trough. Pile smaller pieces on top. Fill with water two thirds of the way to the top (don't overfill the containers with clay, or mixing will be a major mess). Let these sit for a few hours to a few weeks (if soaking for long periods of time, store in the shade or you will end up with a bucket-shaped brick, and you are back to step one!) depending on the clay percentage of your clumps. The lower the clay content, the quicker the clumps will break up. Check frequently. I generally check with a shovel. If it feels like you are hitting a rock, it needs more soaking. You can try to split chunks into smaller pieces in the bucket with the shovel; just be cautious not to break the bucket and spill precious slip everywhere.

I have, on one occasion with some especially tough dry clay, piled it on a tarp, folded the tarp like a burrito and driven over it. It seemed to help, although I certainly wouldn't recommend this technique for a whole house!

If you are in a time crunch, you can set up a ½″ or ¼″ screen over a larger bucket and dump the contents of the soaked buckets (after a bit of agitation to help break up the clumps) on the screen. This lets the already soaked and liquefied clay pass through, and the clump can go back in the bucket with more new clay clumps added in, and the process can be repeated.

Making Slip

Slip is clay particles in suspension. As you are mixing the clay, you may notice that a lighter, creamier color is coming to the surface (as things thicken, you can make beautiful patterns on the top of the container with these two different-colored fluids). This is clay in suspension. As the density of clay in suspension increases, the surface tension of the liquid starts to change. This becomes obvious when something dropped into the mix leaves an indent. The best test for knowing when your slip is ready is to stick your

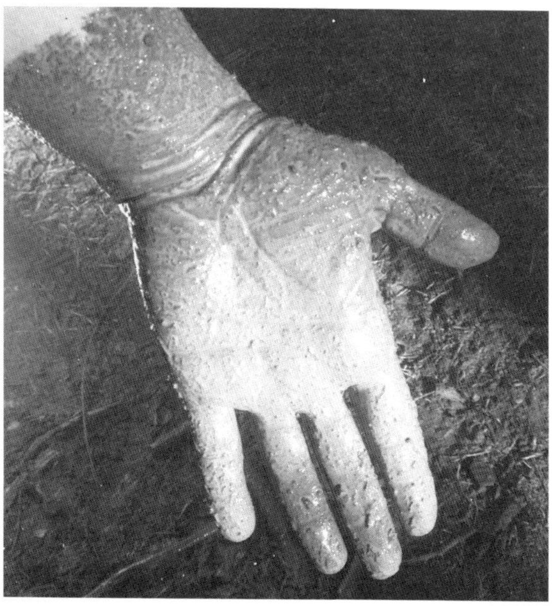

Slip that is too thin to adequately cover straw will run off your hand and reveal your skin below.

Photo credit: Lydia Doleman

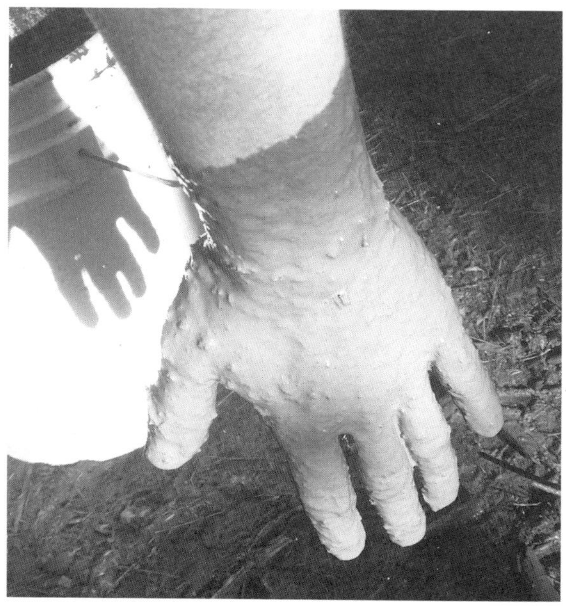

Slip that is too thick will leave webs between fingers and/or look lumpy.

Photo Credit: Lydia Doleman

bare hand into a bucket of potential slip. If, when you pull your hand out, you can no longer see your skin and it doesn't shine through, then your slip is thick enough.

Ideally, slip will be mixed to a consistent formula to achieve a predetermined density in the finished, dried LSC wall. Generally, we are aiming for a dry density of 13 lbs per cubic foot (6.3 lbs of clay and 6.7 lbs of straw) which will yield a thermal resistance of R-1.69/inch.

Details on calculating bagged clay and subsoil quantities for an optimal slip mix were given in Chapter 7.

If, after you remove your hand from the mix, there is a dimple in the mud surface, you may still add water. Don't sweat this too much in the beginning. The spectrum of what is acceptable is pretty broad, and it will be noticeable as the straw clay is being mixed if the slip is too thin; the straw won't mix very well and won't stick together.

The person who mixes the slip is the key player, as without slip all operations cease. If you are planning on sifting your soil as you make slip, plan enough time for that task, and have more people on that end because the slip supply gets used up quickly. Dust masks are a good idea for anyone on the mixing end, as sifting and mixing straw clay are both dusty jobs. Safety glasses make a nice addition; slip or straw in the eye, although not toxic, is also not comfortable.

You have a lot of options in how to go about mixing the clay to get the slip, from manual to industrial.

Mixing by Hand

The most basic and time-consuming method of mixing is to mix by hand — squishing the clay apart and mixing it into a slurry into the water. This is a very tactile and delicious process! However, your hands will get weary of this. Your whole hand will start to feel very raw, which will make you slower in everything else you do. If you aren't into rubber gloves, the nitrile gloves that are mesh on the back are quite comfortable and will save your skin. Leather gloves get heavy and are destroyed in no time by clay.

You can also mix slip in a trough with hoes. Definitely a calorie burner and messy! There are troughs that are curved on the bottom, which are excellent for mixing with hoes; a wheelbarrow is also good. Another option (which also increases the scale) is to make a trough of straw bales lined with a tarp or two to create a giant bowl. Mix by foot, or better yet, boots, as the exfoliation is really only fun for the first 30 minutes.

Simple Mechanical Mixing

The next step up is mixing in 5-gallon buckets with a drill of some sort. Cordless drills tend to wear out and are not meant for this kind of continuous use. A corded drill with a ½" chuck is a good choice for small batches, while a more powerful Hole Hawg drill with a paddle bit is even better; for the committed, the Collomix is a great industrial-strength hand-held mixer for use with 30- and 55-gallon barrels.

If using 5-gallon buckets (I only recommend this for smaller projects because the volume of slip needed requires a large number of buckets of clay slip ready simultaneously to keep up with the straw clay mixing), make a custom cover lid (cut a 1½" × 6" hole in the lid of a 5-gallon bucket) and use a drill or paddle for mixing (a paint mixer bit is not very effective on chunky clay, but will work for dry powdered or sifted clay).

Use the lid on the bucket to keep from getting splattered and getting clay/soil particles in all the crevices of your tools. Depending on the direction of the rotation of the drill, I recommend positioning it so that when you hit the trigger it hits into your hip instead of wanting to spin your arm all the way around (especially if you are using a Hole Hawg). In the event you hit a big chunk or rock, your arm is the weakest link in this operation, and the mixer will spin you around before it stops. Your hipbone is much, much stronger than your wrist (which you will need for all that mixing and tamping). Also, there is a tendency for the bucket with the chunks of clay and the slip to start spinning too — which generally results in people standing on top of buckets with a mixer. Don't do that. It's really not safe. Your best bet is to graduate to a larger container, such as a 30- or 55-gallon drum.

Even for a tiny 10′ × 10′ tool shed, you will use more than 100 gallons of slip. It is best to make big batches and have more time for other things — especially because the person doing the mixing on small projects is the only person who doesn't really get to socialize. A 30-gallon garbage pail is great (though a Hole Hawg with a paddle bit will likely bump any grooves on the bottom of the can) and a 55-gallon drum with an open top is better. Any bigger, and it becomes very difficult to dump out sediments that inevitably build up on the bottom of the mixing barrel. Dump out the barrels when it becomes obvious that the sediments are impeding the mixing process.

Screwing a piece of plywood to the bottom of the bucket so you can stand on it to hold it in place is much safer than trying to mix with one foot on the bucket to hold it steady!
PHOTO CREDIT: JIM RIELAND

Save the sediment, though, because it can be mixed with straw and used in tricky spots on gable ends, missed spots where the straw didn't fill in the form correctly, or random holes. Keep the sludge in a pile or a bin and keep it covered so it doesn't dry out into a brick; that way, it's available when you get to the top of the wall and need straw clay with a higher clay content. You will likely have more of this than you need, so create a plan for what to do with it.

Slip is very heavy, so refrain from filling 5-gallon buckets to the top with slip. Carry two half-full ones at a time and save your back. Or just use smaller buckets to move slip to the mixing station.

A drill press can be modified to work like a giant Hobart mixer, which is excellent because a person doesn't have to stand there while it mixes, and it runs at a slower speed, so less slip is launched out of the mixer.

For most jobs, the Hole Hawg or a Collomix will be adequate to keep up.

Mixing with a Mortar or Concrete Mixer

A mortar mixer can be rented, and it can mix a large volume of slip — as long as the clay you are putting in there doesn't contain too many large rocks (this type of mixer is designed specifically for mixing mortar that has only sand as an aggregate). A concrete mixer will do an okay job with

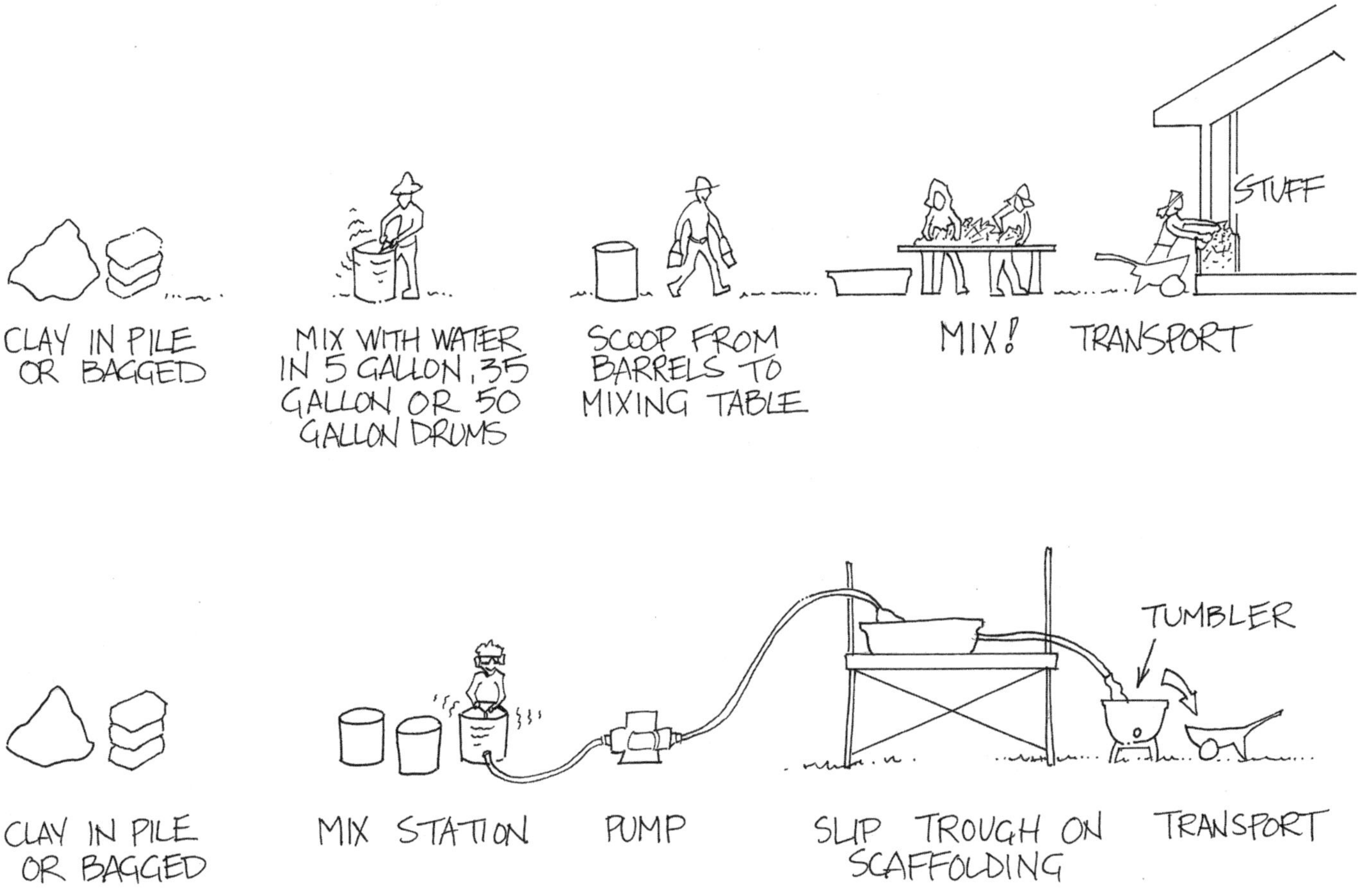

SLIP MIXING OPTIONS

dry clay soil, but not with clumps, and it tends to spurt out a fair amount of the liquid clay.

Protecting Tools

Mud (i.e., water and soil) are the two sworn enemies of electric tools and moving parts. It ruins them quickly! Keep hands clean and dry. Have a bucket of clean water and a towel handy for people using hand mixers, so your tools will survive the job. If you are using temporary or permanent forms, have one person, or a team, of "clean hands" people who move the forms, because they will likely be using cordless drills; if their hands are muddy and they pick up a tool — despite all the admonitions you gave at the beginning of the day about keeping tools clean — they will invariably hasten that tool's demise. Trying to fashion plastic bags around cordless tools doesn't work for long; the best practice is a clean hands person who is in charge of forms.

If the mixing is done with a drill, Hole Hawg, or drill press, make sure you keep a bucket of clean water and a towel nearby; it is inevitable that the mixer will want to test the slip by doing the "glove test." Dirty hands can equal a dead drill by the end of the work day.

Storing Clay, Slip, and Sand

Storing clay appropriately is important. Once again, keep it as close as possible to the mixing site to be efficient and keep it in its chosen state. Buckets of premixed slip will settle out; you'll get water on top, sandy silt on the bottom, and then a layer of clay. If the water evaporates out, you'll have to break it all up again. Store slip in a shady spot if possible, keep it covered, and check it regularly. If you are storing dry clay, lay tarp down and then put plastic and then a tarp on top of the pile. Weigh that cover down so it doesn't blow away. You can also lay down a couple sheets of plywood, which is nice for when you get to the bottom of the pile — you won't be scraping up dirt, grass, and/or gravel. For big projects, where clay is being imported, building a box to store it in is a great idea. The plywood keeps the clay off the ground, and it's easy to shovel or use a front loader to get the clay out. It is also easy to tarp, and you can use the box again for plaster materials.

Storing sand in a plywood bin or just on sheets of plywood keeps sand from disappearing into the ground and keeps it uncontaminated from ground debris.
PHOTO CREDIT: JIM REILAND

A mixing table can accommodate a lot of mixers for quick hand mixing.
PHOTO CREDIT: LYDIA DOLEMAN

Making Light Straw Clay

There are three main ways to make LSC: with a pitchfork, on a table, or with a tumbler.

The "traditional" method was on the ground with pitchforks; however, this method is very hard on people's backs and is less efficient than the table method. The pitchfork method is really only recommended for small projects.

The table method: Mix at kitchen table height, or even a touch higher. Use saw horses and a sheet of plywood. Cover it with a tarp to make a nice smooth surface, and you can have up to eight people all happily mixing — or just two. One person alone is rather slow, but it can be done solo. Dust masks and eye protection are recommended.

The tumbler is a brilliant homemade tool for mixing light straw clay. It is essentially a motorized rotating tube with tines inside to toss the straw and clay slip. An electric motor rotates the tube, which is angled downward, so it is gravity fed. If you have stood around a mix table inhaling dust and getting splattered and loved it, you may very well love the tumbler too! It is a two-person job. One loads the straw and the other unloads the straw clay. The loader and a handful of people un-flake bales and make a massive pile of loose straw by the top end of the tumbler; the group all helps load straw while the loader controls the valve that allows the slip to flow in at the right rate to coat the straw.

A plywood landing area is ideal for receiving the straw clay. From there, it can be easily scooped up by one person and put into bins or buckets and hauled to the work site. A constructed wooden bucket can be made to accommodate forks on a tractor, and that can be loaded.

Slip gets pumped into gravity feed tank, drains down hose and through control valve and into tumbler while straw is added. Photo credit: Lydia Doleman

> To ensure a consistent mix and final LSC density of 13 pounds per cubic foot, remember the basic formula:
>
> 50 gallons of slip contains 4.125 ft^3 of subsoil (6.25 5-gallon buckets *or* five 50-pound bags of bagged clay) and will coat 39.6 ft^3 of straw, which is three 2-string bales.

The Process

Table mixing

Start by cutting the strings on a bale placed at one end of the table and propped vertically. (Tip: I am a stickler for cutting the strings at the knots on bales so you can reuse the strings. Murphy 's Law says that if you leave the knot in the middle of the string you will need to tie a knot just on the wrong side of it later!) Once the bale is open, it will expand a bit and "flakes" can easily be removed from the open bale. A flake is a section of bale that naturally detaches from the rest. The bailing equipment pulls up chunks of straw at a time and these eventually become flakes. Take one or two (more than two takes significantly longer time to thoroughly mix the slip into) and gently shake them out onto the table creating a big pile of loose straw. In the event that you find a moldy spot, now's the time to throw it out.

Pour slip on to the pile. Start with about a gallon on a flake's worth of loose straw (depending on desired consistency), and toss like a salad. Many hands make light work. Within a few minutes of tossing and turning the pile with the slip, the golden straw becomes a dull brown, or grey or red depending on the color of the clay. One tip for mixing is to grab the bottom of the pile where the slip has pooled; visualize cleaning the table with it and turning it over onto the top of the pile. Another effective method is to grab a slip-coated handful and shake it gently out on top of the pile.

In order to determine consistency, take a wad of straw and twist it into a burrito-like shape. If slip comes dripping out of it, you should probably use less slip; if the burrito won't hold together, add more slip or have the slip mixers mix it a little thicker. Once again, the spectrum is very broad for how much slip to add to create a mix that will stick in a removable form system. It is best to do some test patches in the wall. You may be amazed at how little slip is needed to bind all the straw together.

Once the straw and clay has been mixed, shove the mix off the table into a receptacle for someone to take to the building site. Recycling bins, big flexible tubs, boxes, 5-gallon buckets, or even dragging a small tarp are all great receptacles/methods for transport.

It is also possible to make large amounts of slip-covered straw, and let the pile sit for about 24 hours (not much longer than that during hot weather). The water soaks into the straw, making the mix a bit stickier. If you have limited number of people, a lot of premixing could happen on the day prior to a bigger work party.

Twist the LSC to test for drips. If it drips, it is too clayey. It if twists and sticks, it is just right. If it doesn't stick — add more slip.

Photo credit: Lydia Doleman

Tumbler mixing

For larger-scale projects that are implementing more equipment (tractors, pumps, and tumblers) follow these steps:

Have clay prepped, and identify a good dumping spot for unwanted sediments like organic matter and/or bigger aggregates, or a place ready for empty bags from bagged clay. Have a designated transporter, be it a tractor driver or wheelbarrow operator, or create a snag-free runway to drag a tarp-load of LSC into the building.

The slip mixer will need:

- at least five 55-gallon drums with the 50-gallon level marked
- power source for mixer
- water
- borax and measuring cup (if using borax or other water thinner)
- shovel for scraping barrels
- pump (and electricity or fuel for pump)
- personal protective equipment (glasses, dust mask, hearing protection, and gloves)
- small screen if removing organic matter that floats on top of slip-filled barrels

Start by filling barrels with the determined amount of water (based on your prior test batch). It is much easier to add dry materials *to* water than to stir up dry materials that have been compacted to the bottom. Mark your water fill line (spray paint works well). If you are adding borax or other water thinner, now is the time (before the soil is added), then add the specified amount of clay/soil. Mix with the Hole Hawg or the Collomix. Once again, be mindful not to overfill the barrel, or much of the slip will end up on the person doing the mixing. Waste not, want not!

After the mix is brought to the desired consistency (chocolate milkshake), skim off any unwanted organic matter that has surfaced (not necessary if you are using bagged clay or very finely pre-sifted clay soil). This isn't a vital step, but it's helpful if you are using a pump to a gravity feed tank. Once all barrels are mixed, turn the pump on and send it to the gravity tank or trough. If there is more than two inches of sediment on the bottom of the barrels, remove it. (Save it or pile it somewhere, because you might want to use it later to fill holes and do repairs. But if the sediment won't form the shape of a ball, don't bother saving it. It won't be useful.) If you let the sediment build up in the bottom of the barrel, it will ruin the ratios in your mix (if you're using fill lines marked in your barrels).

The slip trough

This needs some steady mixing to keep all the clay particles in suspension — unless you have an endless stream of slip entering the trough. Have one clay-loving fiend up in the trough walking barefoot in circles to keep it in suspension. It's a tough job, but somebody's got to do it. If the weather is cold or no one wants clay between their toes, a person can stir with a hoe or a mixer to keep the particles in suspension. The slip trough needs to be higher than the tumbler so it can gravity feed into the tumbler. A flight of scaffolding should be sufficient. Too high, though, and it could get top heavy. Troughs can be galvanized horse troughs or plastic water troughs. Generally, it makes sense to have 100 gallons or more of slip when using a tumbler. From the trough, run a 4″ hose/pipe to the tank. A clear hose is helpful in identifying the location of the clogs that have the tendency to develop with a silty slip mix or when there is only a very gentle slope to the tumbler. The plumbing rule of thumb is to have a ¼″ rise per foot to help flush solids along; this is a good starting slope for the pipe that runs from the trough to the tumbler.

A fun way to keep the clay particles in suspension!
PHOTO CREDIT: LYDIA DOLEMAN

Tamping

Tamping the light straw clay is very simple — but proper tamping can make or break a project.

Consistency is key, so it is essential that the whole crew (paid or volunteer) be knowledgeable *right from the start* about how to tamp. It is wise to do a demo. Repeat it each time someone new comes on to do tamping.

Tamping is incredibly simple, quiet, and easy: Take a handful or two at a time (the narrower the wall, the less you put in at a time) and place it in the wall cavity. Tamp with a tamper. Tampers can be scrap wood from on site, like scrap 2×4s or 2×2s, or custom made to fit in a person's hand nicely. The people doing the tamping should focus on the corners and edges with temporary formwork, placing one third of their body weight onto the tamper to compact the LSC into the corners. For permanent formwork, the tamping is lighter, so as to not blow out the mesh, reed mat, bamboo, or wood lath.

A variety of tampers.
PHOTO CREDIT: LYDIA DOLEMAN

Starting with smaller handfuls until each person is accustomed to how much pressure to use and then removing the form immediately after reaching the top allows workers to get immediate feedback on the amount of pressure they are using and whether they are cramming in the right amount of LSC. With thick walls, people can walk in the wall cavity to help tamp down the LSC as well.

Things to consider while tamping: careful packing around electrical boxes and wires, horizontal supports, and making sure to weave the straw through Larsen truss cavities and in stud bay keys. Hazards include nails that might stick through framing, splinters, and hitting electrical wires.

How do you know if you tamped it right?

Before you start filling in walls, you should make several test blocks. For one, you should tamp, tamp, tamp really hard; for another test block, compress the LSC with about one third your body's weight; for the third, gingerly pack it in. Remove your form, and you will see significant differences. The over-tamped block will have highly compressed straw, few voids, and will be significantly heavier than the others. It will not give when you press on it, and will have a very dull thud if you tap on it. At the other end of the spectrum, the test block that was under-tamped will be light and fluffy, probably have pockets, and will feel very spongy to the touch. This one won't really generate a sound if tapped, and will give when pressed. This is not ideal: under-tamping makes for a very poor plaster substrate. You are looking for the middle ground: firm but not so compressed that you have obliterated the voids the straw contains in its tubular structure. When you pull the form off, there are no straws poking out, and there are no pockets.

When working on the main project, pull forms off immediately so workers on the wall can see their work and learn quickly. If a section appears under-tamped, pull it out. It will not hold plaster well and will be a long-term problem. Again, you should instruct your helpers to focus on tamping corners and edges, putting a third of their body weight on the tamper until they get the hang of it. The body of the wall cavity will be held in place by the corners and the edges being tamped well.

It is much easier to fill holes before the wall system dries completely. A quick method is to have a bucket of slip available and to dip a small amount of LSC in the bucket and to fill voids as forms are removed. Use a trowel or a small section of plywood to flatten out the patch so it stays in the same plane as the rest of the LSC.

Challenging spots

The top of the wall is usually a spot that causes a fair amount of head scratching — unless you have the roof open and you can tamp entirely from above. For these odd sections (also found below windows and in gable ends), it is best to create micro forms from scraps of plywood and do the tamping from the side. For the very last section, where you are pretty much tamping at a 90-degree angle from the studs or trusses, dip the LSC into a bucket of slip to make it stickier, and, pressing firmly with hands, give it a "comb over" where you slide your hand over the loose straw over and away from the stud toward the body of the wall cavity to slick the LSC over and help it adhere to the rest of the LSC in the stud bay. It is helpful to flatten it out with a trowel or compress the form over the side-tamped area to keep it flat.

The same method can be applied on gable ends. Usually, the form is removed and angled parallel to the roof joists, and then the same top-of-the-wall technique is applied.

Wall section, correctly tamped. Note even packing, with no loose straws or pockets. Photo credit: Lydia Doleman

Notice the shaggy look and the voids in the wall panels on the left. These portions of the wall have been under-tamped and run the risk of being a poor plaster substrate. Photo credit: Lydia Doleman

Note the darker mix in the top 8″ of wall section. A heavier-clay LSC was applied and packed in to help it adhere with side tamping. It was then scraped flat. PHOTO CREDIT: ERICA ANN BUSH

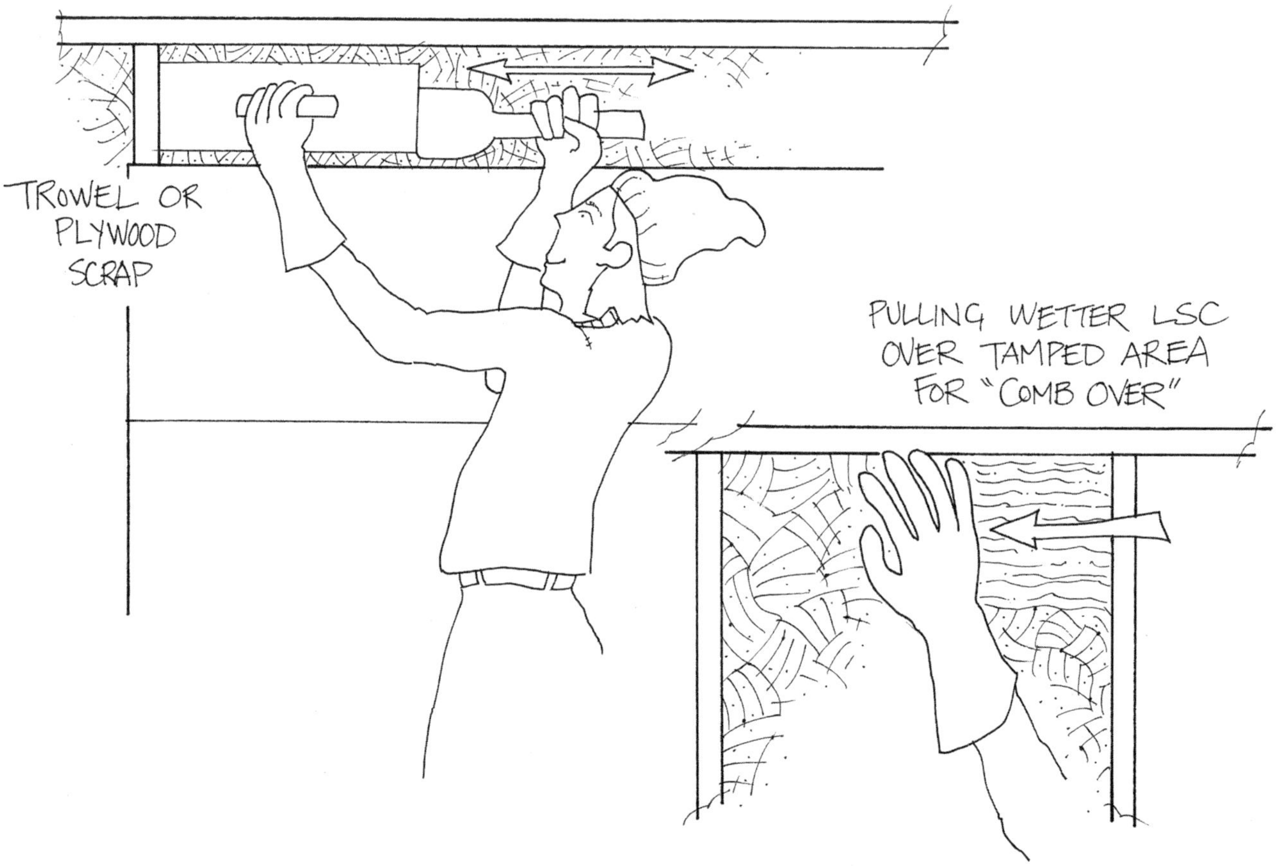

Temporary Formwork

With the strongback system, you are starting at the bottom of the wall and filling in the first section of the wall to the top of the form height. Pull the form off to check your work, make any repairs necessary, and then put the original form back on and stack the second form on top, into the groove. Be sure the forms are fastened flat to the studs or trusses, because a bulge will impact the flatness of the wall and will either have to be ground down later with a grinder (a dusty affair!) or will leave a lump in the wall.

Remember to place horizontal reinforcing rods. These are easy to forget. These are ridged 2″ or less in diameter lengths of material. Most commonly used are wood and bamboo, followed by wire, poly twine, or string. A good way to stay on top of putting these in the wall is to use brightly colored spray paint at the prescribed heights for placing them in the wall cavities as a reminder. If you have a "dry hands" person dedicated to moving the forms, that person can also be in charge of the reinforcing rods.

Permanent Formwork

Reed mat

For this, you will need to have one person be the roller and stretcher and one person be the stapler. Starting from the bottom, hold the reed mat parallel to the floor and fasten it to bottom plate. Stapling through baling twine, wire, poly mesh, or rope can keep the staples from blowing through the reed. Once the bottom is stapled, the reed mat can be rolled up 16″ to 24″ and stapled along the studs. Care should be taken to roll tightly so there is little bulging of the mat.

Packing in the reed mat requires less pressure than with temporary forms. However a balance must be struck, because you don't want to under-pack. Holding a piece of plywood outside of the reed mat (a plasterer's *hawk* is nice) while compressing the straw inside can help prevent bowing while packing.

Once again, you want to pack the wall full enough to be firm to the touch, but not so light that the wall gives when you press on it, as this will eventually lead to plaster cracks.

Wood lath formwork

Wood lath can be stapled on as you go, working in lifts from 16″ to 24″ at a time. The lath is more resistant to bowing, so you can pack more in. It is also very forgiving of under-packing, as the wood has the strength to carry the plaster and doesn't need the LSC to keep its form.

Lath is traditionally nailed on, but a pneumatic stapler is far more efficient. One tip to fastening with a stapler is to staple diagonally over the grain — not horizontally, with the grain. This creates two points of contact to help hold the wood in place. This is helpful when using dry lath, which has a tendency to split when you are stapling at the very ends of the pieces of wood. The advantage of the wood lath is that it is relatively rigid and can hold a denser mix; it also provides a rigid substrate for the plaster. A soft substrate over a minimally packed infill will lend itself to cracks — a very challenging situation to remedy after the fact.

Wood lath is best applied completely from bottom plate to top plate on one side and then installed as the light straw clay is being installed on the other, much like formwork. Having one person on a crew dedicated to installing the lath with clean, dry hands will make for a much more efficient project and longer-lasting tools.

Wood lath can be used in curving wall systems and curving window openings, especially above windows and doors. In cases where LSC is being used to insulate ceiling joists, wood lath can be bent in place to create a convex plaster substrate between the joists. It is a fair amount of labor, but makes for a unique aesthetic.

Wire mesh/metal lath

Wire mesh/lath is typically stapled in place. Because wire mesh is flexible, it is well suited for curving wall sections. Depending on the type of wire, using mesh/lath can result in spongy walls, so may not be appropriate for 48″ stud spacing, or it may need to be doubled up (chicken wire with 4×4 cattle panels can be used).

Wire mesh is stapled into place — with care taken to keep it in line with the floor and ceiling and to keeping it taut. To add rigidity to the mesh and make for a better plaster substrate, create leverage while installing the wire mesh by using a rake or 2× material to pull on the mesh. If the mesh becomes saggy, there will likely be too much flexibility in the substrate for the plaster and cracks will form, or a bulge might occur with too much LSC packed behind loose mesh. Wire mesh can be a good, fast, and easily attainable alternative to reed mat or wood lath, especially if used on the exterior where a rainscreen is to be installed.

Split bamboo mat

Split bamboo mat can also be used as permanent formwork. See Chapter 6, under "Permanent Formwork."

Drying

The rule of thumb is that LSC dries at a rate of one inch per week from both sides of the wall in a warm climate, faster in windy conditions, and slower in cool, humid conditions. Due to the extended drying time, especially in cool climates, aim to finish infilling with sufficient weeks of summer for the walls to dry, or use fans to speed the process.

During drying, little white fuzzy spots of efflorescence are normal, as are green sprouts from seeds in the mix, which will wither and die as the wall dries. Adding borax to the slip can help discourage mold development by both changing the pH and also diminishing the amount of water in the mix, and therefore reducing the dry time. White, and usually green, molds are harmless. Black mold on the light straw clay is undesirable and potentially harmful; it can indicate that the walls are not drying fast enough. A fan and dehumidifier should be implemented immediately.

The moisture load a light straw clay installation can put on a building can induce mold spots on green wood. Ventilation is key for expediting the drying process and not adding moisture stress to the rest of the project.

Workshops, Work Parties, and/or Work Force?

To workshop or to work party, that is the question!

There is this notion that if you host a workshop to build your house, people will pay *you* to build *your* house, so you get something for free. This is partially true. Yes, you will receive some free labor and yes, people will pay to work on your project. However, it can't be stressed enough that a workshop is not something for nothing. People pay money (sometimes lots of money) to take a workshop *to learn how to build.* The emphasis is learning, not necessarily productivity. People who pay for something generally have expectations, making the workshop more about them and less about the project. On top of that, people who are in the process of learning how to do something will be slower than someone with experience, and they will probably make some mistakes. God forbid you get a group of fast-moving mistake makers! Those mistakes will be on *your* building and might be costly to repair. Also, workshops have the inherent risk that they might not fill up, leaving you without a labor pool and an instructor without a job to do after many hour spent promoting it.

Workshops are an awesome gift to give to people who want to learn from people who are very interested in educating. If you have taken a workshop yourself, have an idea of who is expert enough to teach one, and think you can attract enough people to attend one, then it may well be an excellent fit for your project. If not, you might want to consider a work party or hiring a crew.

A work party is just that: a group of folks coming together in a lighthearted atmosphere to accomplish a task. Food and water should be provided, along with a fun atmosphere: music, beer, etc. (Although, productivity will plummet if you bring the beer out too early. The same goes for the food.) If folks feel like they had a good time, were useful, met good people, and were well fed, they are likely to make themselves available again — which will be especially handy when it comes time to plaster. If they were hungry, felt like they stood around with nothing to do, were treated like a slave, or generally didn't have a good time, you might be watching the tumbleweeds roll by when the next work party happens.

One of the nicest things about straw clay is that it is a quiet activity that can be highly social; on the downside, it can be overwhelming for one person to tackle a large project, so making it enjoyable and efficient is imperative, especially for big projects — and a typical residence of over 1,500 square feet is a big project.

As mentioned before, a minimum of five people are needed for a smooth installation. The primary person is the slip mixer. For a very basic project, that person (or handful of people) mixes slip in a bucket, a trough, a wheelbarrow, or 55-gallon barrels and operates the pump to the gravity feed trough for the tumbler. This person is versed in what the desired consistency is. For most projects, the spectrum of what is usable is broad. Slip can be mixed to a push-the-limits thinness, or it can be a thick, chunky pancake-batter consistency. Consistency is the key and that involves keeping a good track on amount of water to clay ratios. For small projects this can be a "to taste" sort of process. One person adds water to the mix, another wheelbarrows over sifted soil, another operates the mixer. Slip is then either moved to the mixing table or pumped up to a trough. For the mixing table, it is best to have a 5-gallon container or something larger that the slip makers just dump into because straw gets in, on, and all over everything by the table. Then the mixers can dip a smaller bucket in to pour on the straw.

It is especially important for you to have a steady stream of things for volunteers to do. Figure out your systems and work stations ahead of the workday, prior to volunteers arriving. If at all possible, prepare a good amount of slip the day before. You will probably need some friends or volunteers you can count on to help with this "pre-work party" preparation. What you are working to avoid are the two common bottlenecks: the slip mixing and the straw clay mixing. Actually putting the LSC in the walls usually goes rather quickly, and most people gravitate to stuffing the wall; you'll want to direct some of those folks to swing into mixing the straw clay. But it's the slip making that is really key; if you run behind on slip making, everything will come to a grinding halt. Having a lot of slip made *prior* to the work party helps keep things running smoothly, especially if you encounter a bump in the program.

There is something to be said for hiring committed people — people who agree to be present at a certain time for a certain amount of time with a certain amount of productivity expected from them for a certain price. It is nice to have a level of reliability and responsibility to count on — something you are not guaranteed to get from a workshop or work party volunteers. I highly recommend hiring at least one person, especially someone committed to mixing slip for a work party, or someone who has construction experience and understands how an efficient work site flows; you will get very busy very quickly greeting people, answering questions, giving directions to the bathroom, etc.

Depending on your budget and timeframe, you might want to hire a few people:

- one person dedicated to slip mixing
- one person dedicated to the tumbler or the mixing table
- one person on quality control in the house (this could also be your dry hands formworker)
- one person leading the crew on tamping
- one person to operate heavy equipment if you are moving your LSC via a tractor
- one person to handle carpentry issues like forgotten blocking, or form building, or building ramps or anything that can make the job go smoother

A dream team for a work party would include:

- one person just mixing slip
- if you are using a tumbler, one person who's sole job is to run the tumbler
- one person to guide the mixing of the straw clay
- one team leader on the inside of the house directing the wall filling
- someone to answer general questions
- someone to handle food and drink
- someone to stick around and help you clean up at the end of the day

How much can you do in a day?

Using a table for mixing, a crew of 8 (inexperienced, but good workers) can fill 160 cubic feet in two 8-hour days, or 20 linear feet of 8′ tall wall that is 12″ thick.

With a tumbler, a mixing team of three and 12–15 stuffers, 900 cubic feet can be filled in three days, or 112.5 linear feet of 8′ high by 12″ thick wall.

There are lots of considerations for predicting how fast a project will go:

- Complicated wall systems take more time
- Moving forms on both interior and exterior
- Tall walls
- Distance from mixing site to site
- Weather
- General speed of crew/level of experience
- Accidents/repairs

Many hands make light work!
PHOTO CREDIT: LYDIA DOLEMAN

Chapter 9
Finishes

One of the most appealing features of LSC is its ability to be finished in a variety of ways: wood cladding, siding, earth plaster, lime plaster, gypsum plaster, or sheet rock. It's an unusual construction type in that the structural framing itself is present in the surface of the wall. This allows for mounting various styles of siding; in addition, LSC is a flat wall system with a lot of surface area to hold plasters well.

The most common and traditional finish is a plaster finish, although in very wind/rain-prone areas, a rainscreen of either half height or

The south face of this LSC brewery is protected with a full rainscreen on the south side of the building, where the weather hits hardest. Photo credit: Dean Hawn, Burning Daylight Construction

Earthen interior plaster of a Laporte EcoNest. American Clay is the company that manufactures this beautiful plaster. Photo credit: Lydia Doleman

Lime-plastered exterior of a Laporte EcoNest. Photo credit: Lydia Doleman

full height is an appropriate finish. Each finish has pros and cons and should be decided on well in advance of construction. See Chapter 5, Material Specifications, for more information on plasters.

There are a variety of plaster styles, techniques, and finishes. Earthen plaster is the most basic and versatile. Earth is breathable, infinitely recyclable, and repairable; it can be applied in a wide range of colors and textures, and it provides excellent humidity regulation of interior air and for structural wood (because mold isn't encouraged beneath a plaster skin). Earth is very user friendly and economical. You can make your own, or you can purchase premade and mixed earthen plasters from companies like American Clay. You might want to purchase premixed plaster only for your finish layer.

The one disadvantage of an earthen plaster is that it is water soluble. When using earthen plaster in a kitchen or bathroom, it is important to incorporate strategies for keeping water from directly dripping on or spraying the walls. In areas where there is a high probability of water contact or a desire for washability, lime is the better choice.

Plaster Prep

Taping

The human eye is very keen at noticing lines. Taping is an essential part of plastering in that it helps keep a clean edge with the plaster as it tucks behind or ends into trims and frames. One can plaster without tape, but it is a messy affair, and the cleanup can seem interminable because all plaster binders have a way of working themselves into wood and other materials. It is also very hard to clean right up to an edge without hitting the plaster.

Start taping from the bottom up leaving the exact distance you anticipate the plaster being from the wall with the tape. Tape from the floor or baseboard, up. Tape any surface the plaster touches: trim, windows, baseboards, electrical boxes, exposed timber frames, and ceilings. Typically, it is ¼″ for the first coat or rough coat, ⅛″ to ¼″ for the second, or brown, coat and 1⁄16″ to ⅛″ for the finish, or color, coat. Tape is usually removed from the top down, so applying it bottom up helps in the removal of the tape by keeping a taped surface at the bottom to catch all the plaster that might crumble off from the tape above. Use good tape. Blue tape/painters tape is good for plasters, but when misted heavily, it tends to bubble. Vinyl tape is resistant to water, but doesn't stick to all surfaces and may leave a residue. Test the tape on wood that has been finished, as some finishes resist the adhesion of tape. Make sure to test the tape under misted conditions and let it stay on the substrate for 24 hours to see if it leaves a residue.

Tape and paper the ceilings, as plasterers will invariably leave knuckle marks on the ceiling. When using J-channel, tape under the J-channel, because some have weep holes and will leak moisture — which could potentially stain flooring and be very difficult to clean up. There is no need to tape the 3⁄32″ metal reveal on the J-channel because it wipes clean after the earth or lime plaster is dry. Tape and paper exposed timber and pole frames and use plastic and tape to protect windows, sills, and other horizontal surfaces. It is wise to lay down dropcloths with plastic underneath them if working over finished floors.

Doors, doorknobs, and door openings will benefit from being papered and taped — plaster-covered hands on the mechanics of the doorknob will do some hard-to-repair damage, and it is all too easy to bonk a door or a door jam on the way though with buckets or wheelbarrows of plaster. Also be sure to protect stairs and thresholds. The sand in plaster can mar a wood surface if stepped on repeatedly. Cardboard is a good floor and stair protector.

While this may seem like an excessive amount of tape and paper, it really saves on cleaning and limits damage — both lime and clay particles are very small and can get into wood grain or stain materials permanently. Also, be mindful of the mixing area's proximity to the house. If the exterior is already done, stay far enough away to avoid errant splashes on the exterior or on parked cars.

Prep for Earthen Plasters

Wood expands and contracts at different rates than LSC does. Therefore, prep for earthen plaster means carefully bridging the areas between Larsen trusses (or studs or split studs) that will be covered with plaster. There are a variety of ways to do this. The most basic is to use a very straw-rich first *scratch coat* of plaster. This mix is often about 60% clay and 40% chopped straw. It is spread over the wall and over the studs and relies on the built-in tensile strength of the straw in the clay scratch coat to mitigate any cracks that might telegraph through as the wood expands and contracts at a different rate. In order to increase adhesion of the scratch coat to the stud face, a wheat paste/sand mix can be painted over the studs first.

To make wheat paste, stir 1 cup of white flour into 1 cup of cold water. Once this mixture is no longer lumpy, it can be mixed into 4 cups of boiling water by slowly pouring it in and stirring. Run this through a 1⁄16″ sieve or screen, and it can then be mixed with 3 cups of fine sand and painted on the studs. Scale up to get 5 gallons of mix for your average-size residence. Five gallons will keep for a week in cool temperatures, but in the event it starts to go rancid, make more (you'll know it's rancid the minute the mixture starts to smell bad). This technique can also be used for covering plywood or other materials. For larger surface areas, burlap is recommended.

The next step in prepping for plaster is either stapling strips of burlap over the studs/exposed wood members or dipping the burlap in a thick slip (melted ice cream consistency) and adhering it to the studs via the sticky clay in the slip. This process is very much like papier-maché. Care needs to be taken to flatten out the burlap

Burlap being slipped and spread onto studs.
PHOTO CREDIT: LYDIA DOLEMAN

so that it doesn't dry with upraised edges, as those will make the next coat of plaster more challenging to keep in one plane.

Burlap comes in a variety of weaves; the tighter the weave, the harder it is to get clay slip to impregnate each of the tiny openings. It may take a bit more slip and more force if the tight-weave burlap is first stapled onto the studs instead of being dipped. The dipping/papier-maché method is messy, but it ensures that both sides of the burlap are saturated with clay slip and therefore will stick solidly to the stud and straw clay surfaces.

A more (embodied) energy-intensive way to bridge this gap is to use mesh and a hammer tacker or stapler (see Chapter 6, "Design Options," for more on meshes). The mesh tape should be wider than the stud or wood member by at least 1″ on either side. Any mesh can be used as long as it is affixed to the studs, lays flat, and the subsequent plaster can bond to it mechanically. With metal mesh, care should be taken to hammer down or flatten sharp raised edges to prevent hand injuries during subsequent applications of plaster.

Acceptable materials to bridge over studs:

- wheat paste prior to a straw-rich first coat of plaster
- burlap
- ⅛″–¾″ fabric mesh
- fiberglass mesh tape
- expanded metal mesh (chicken wire or ¼″–1″ hardware cloth)

Prep for Lime Plasters

Some walls are plastered with an earthen scratch coat and then the amount of lime is increased in each layer until the final layer is a pure lime/sand layer. The advantages to this are that you are using more of a material with less embodied energy (earth) and limiting the use of a higher embodied-energy material (lime). Downsides are that bonding between dissimilar layers can be difficult to achieve, and care must be taken to make sure that each layer bonds both mechanically and materially to each other. This is achieved by firmly pressing the plaster into the straw clay. Use a small trowel or your hand to push the plaster into the small spaces exposed on the face of the LSC wall. Then flatten each area out with a larger trowel (or plasterer's hawk).

Your plan might call for using *stucco paper* (it's like tar paper and sometimes comes in a two-layer role) over the studs if lime will be used for all three layers. This papering of the studs should be lapped in a way that will discourage water from running behind the paper. Start at the bottom and work up, layering as necessary as you go. The paper can be a part of your airfin system at the tops, bottoms, and corners of walls to control air/vapor movement through the wall assembly. Burlap (ideally a wide-open weave) can be stapled over the stucco paper; fiberglass or metal mesh can also be used over studs, or on plywood.

Plaster is generally applied in three coats: the scratch (rough) coat, the brown coat, and the finish coat. This is true for earthen, lime, and gypsum plasters (discussed in Chapter 5). Sometimes a two-coat system can be used: a thicker first coat followed by a final coat. This is a possibility because LSC walls are (can be) so flat.

Applying an Earthen Plaster Scratch/Rough Coat

With an earthen plaster, the scratch coat is usually a high-clay-content plaster with a lot of fiber. The fiber in an earthen plaster is usually straw or horse or cow manure. This coat keys into the

LSC and is very sticky. It is important in this phase to work out any lumps or bumps in the wall, as they will telegraph through all succeeding layers of plaster.

These ratios can be applied to site clay, though it is very important to make some plaster samples first to ensure that your first coat of plaster is quality: no major cracking issues (too high clay content) nor too crumbly (too much sand).

The scratch coat is best applied with short trowels to really push it into the wall to give it a solid mechanical bond; this is important because all other layers will hang on this layer. Mist the wall prior to applying the plaster as this helps the first coat adhere better. You can expect to find many small cracks in this layer when it dries. If the cracks are $\frac{1}{4}$″ thick (or larger), this is an indication of a problem: an expansive clay was used; there is something flexing in the substrate; insufficient straw was used; or the plasterer didn't integrate one trowel's worth of plaster very well into the next. If you have issues like these, remove the failing plaster and get to the root of the problem. If the substrate is soft, add a ridged metal mesh to help stiffen it up. If the substrate is not moving, add more straw, or try to rework the mix and reapply making sure to moisten the substrate, because a lack of moisture on the primary layer can cause the top layer to dry out too quickly and adhere poorly.

This layer should get floated with a large float to remove high spots and reveal where low spots need to be filled (fill these now). Check for level with a straight stick that is the height of the wall (also known as a *truth stick*). Any irregularities in this layer will have to be worked out in the next. Use a long float, and, as the wall is getting dryer, use either a toothed trowel or the edge of a regular trowel to score the wall to add key for the next layer.

Earthen Scratch Coat Ratios:

1 part clay to 1 or 1.5 parts fine sand and 0.5–1 parts fiber.

Lime plaster, scratched and ready for next layer of plaster.
PHOTO CREDIT: JIM RIELAND

A selection of trowels used for both earthen and lime plasters. Note the variety of sizes and shapes. Not shown is a toothed trowel, which is handy for scratching the first plaster layer.
PHOTO CREDIT: JIM RIELAND

Applying Lime Plasters

Lime Scratch Coat Ratios

The scratch coat for lime plaster is usually a high-lime mix of 1 part lime to 1.5 to 2 parts fine sand. Lime may be in putty or powder form. Commonly, fiber is added at a ratio of ½ part or 1 part depending on preference.

Fibers used in lime and earthen plasters are straw, manure, animal hair, or nylon. With a lime plaster, always mist the straw clay wall to help aid in creating good "suction." The added water allows the lime plaster to maintain its water content longer, allowing it to "cure" rather than dry, like an earthen plaster. If lime plaster is troweled onto a dry LSC wall system, the clay in the light straw clay would immediately soaks up the water from the lime plaster, causing it to dry and break the microscopic bond between the curing lime and the hydrated clay at the surface, robbing the lime of the H_2O it needs for its curing process.

It is important between coats to make sure there is the possibility of a mechanical bond. Scratching between coats gives plaster a "key" to hold onto for the mechanical bond.

Misting between coats is very important to ensure a good bond between layers. The previous coat should not be dripping, but it should be visibly darkened by moisture. One way of ensuring a good level of moisture in the previous layer is to dampen it with a garden hose set on "mist" or use a pump sprayer. Mist till shiny, but not dripping. Let the water soak in until the plaster appears dull again. Do this three times. On hot, windy days this needs to be redone frequently, and lime-plastered walls need to be kept in the shade. Earthen plasters benefit from shade, but direct sun and drying don't always compromise their structural integrity.

Plasters (both earthen and lime) are best if applied in a top-down manner. It is truly only necessary for the finish coat, but it is a good habit to get into. This is important because gravity will want to pull plaster down in drips and drops; you don't want to perfect one section and then drip onto it while plastering above it. That just creates more work. Plastering contiguous areas is also good practice because you don't want separate plastered sections to dry at their edges and create seams in the middle of a wall. If one person starts at the wall and gets ahead, then the next person can plaster in the same direction but a little bit lower, with the third and/or fourth person plastering down to the bottom of the wall. Keeping a wet edge prevents vertical or horizontal seams from appearing, should all the workers stop at once.

The second coat, or *brown coat,* is applied in the same manner. This coat has a higher sand content and is the last chance to work out any lumps in the wall.

For a flat wall, use the truth stick (a straight board that is as tall as the wall that can be held up to the wall to screen for low or high spots) and use long, ridged trowels. Float the wall with a float that is as long as possible, to flatten everything out. (A float is a wooden trowel [usually] that is used like a straight edge on the wall.) Dragging it gently across the surface flattens out high spots and highlights low spots to be filled while pulling the sand to the surface. The process of floating should leave a surface rough

Second/Brown Coat Plaster Ratios

For an earthen plaster, the mix is generally 1 part clay to 2 to 2.5 parts fine sand and 0.5 parts fiber.

For a lime plaster, the mix is also generally 1 part lime powder or putty to 2–2.5 parts fine sand and 0.5 parts fiber.

enough for the final coat. If the final coat is to be thicker than ¼″ of plaster, though, score the wall lightly with a toothed trowel.

Once the second coat has dried and all finish carpentry has been installed, it is time for the finish coat. At this point, a lot of finish work has been done in the building: floors, plumbing fixtures, etc. are in, so it is important to protect those surfaces from the finish plasters. Remember that plasters have sand in them — any drips and drops, when stepped on, will mar most surfaces; lime plasters can and will stain wood, and plaster in general leaves a residue when splattered or troweled up to an edge. Good taping and protection are key to a good-looking finish plaster and avoiding the need for any repair work to carpentry.

A good trim detail is having a beveled edge that plaster can be tucked behind to make the finish plaster look good and to hide the minor shrinkage gap you get at the edge of plasters where they meet other materials.

Happy plasterers using a float and a truth stick to create a near-perfect wall surface. Note the marks on the right that indicate low spots. Photo credit: Jim Reiland

Plaster-to-trim details

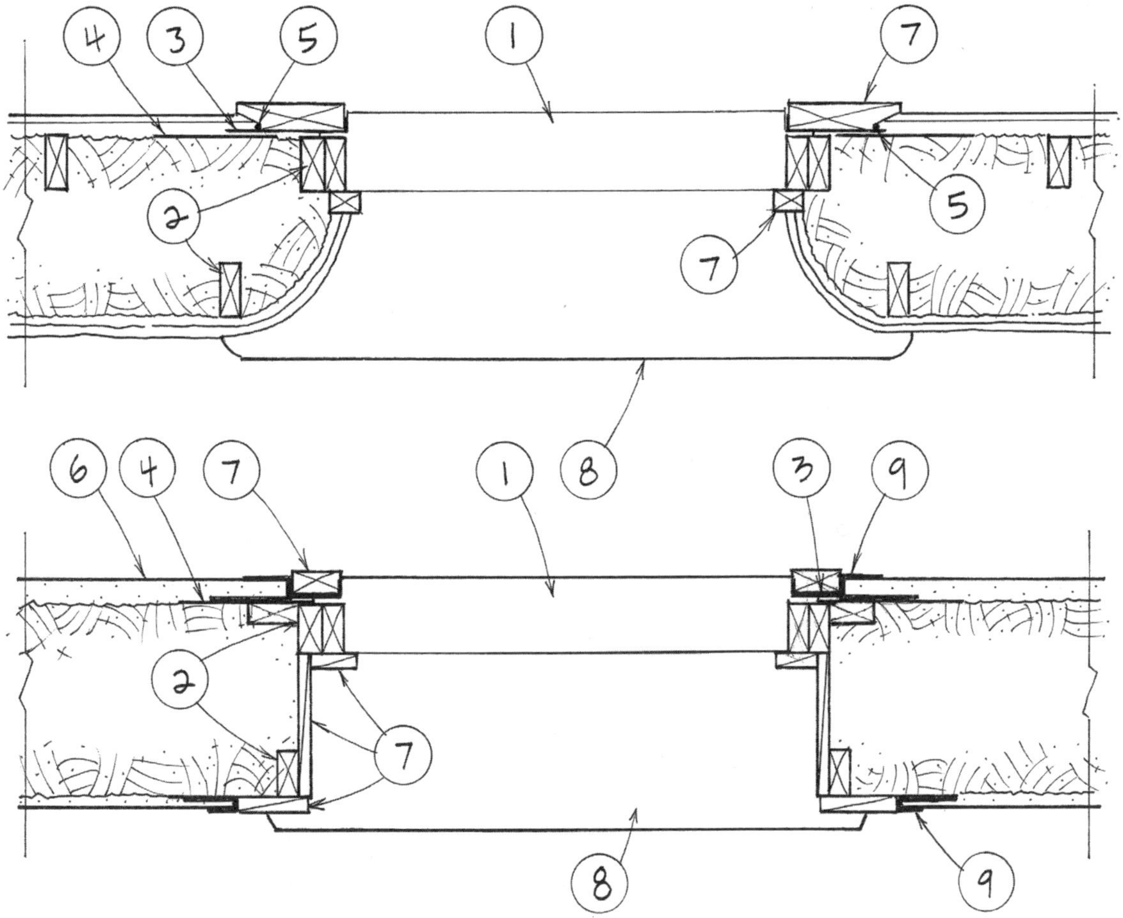

1. Window
2. Framing
3. Ice and water shield
4. Building paper
5. Bead of caulk
6. 3-coat plaster
7. Trim
8. Sill
9. Channel

Finish Coat

The finish coat recipe is generally the same as for the second coat, but with finer sand and a light-colored clay. Fiber is generally omitted from this layer, except for aesthetics (when in doubt, use cattail fiber or nylon, as these fibers are not noticeable in the final coat). The final coat is the layer to which pigment is added.

Finish Coat Ratios

1 part clay (bagged or site clay) to 2–2.5 parts fine sand. Pigment and fiber to taste.

A test batch (or batches) must be made to extrapolate how much pigment to put in the final coat. This is important because pigments are added wet, meaning you must mix water with your dry powdered pigments and then add them to your plaster or add them to your mixing water. If you do not know how much pigment and/or water to use, you run the risk of overhydrating your plaster.

Saving Clay Plaster for Future Repairs

It is extremely helpful to dry some clay plaster into "cookies" that can be rehydrated for future repairs, thus eliminating the need to color match.

Just as for the previous coats, starting at the top of the wall and keeping a wet edge are important to getting a good finish. Depending on the desired texture, a finish plaster can be marbleized (where colors are mixed while on the wall) and extremely flat and smooth (although the more polished a plaster, the less breathable it is) or rough and rustic. With an earthen plaster, a good trick for unifying the texture is to sponge the whole thing after it had dried completely; this can remove all trowel marks. The same can be done with a lime finish; however, it has to be done *before* the last layer is completely dry (when it is at what is called the "green hard" stage, where it resists a thumb print but not a thumb nail). Sponging can also happen after the finish plaster is dried. This will erase most of the trowel marks, but not all. After sponging a dry wall, the wall will need to be dusted of the sand that is loosely clinging to the surface so the wall doesn't appear to be dusting.

Pigments are typically integrated into plasters during the mixing stage, but is possible to spray on the pigment as the lime cures (or on an earthen plaster) and trowel it in for a mottled look or for very vibrant colors. The sky is the limit when it comes to artistic plaster finishes!

Chapter 10

Maintenance and Renovation

A GOOD DESIGN AND A GOOD ROOF are basically all you need to maintain a light straw clay wall. LSC buildings that have never been plastered have survived the elements for years; and there are buildings in Europe that have sustained significant damage to their plasters, and the straw clay has remained intact for decades! That being said, the enemies of LSC walls are point loads of moisture (where moisture is concentrated into one spot, e.g., a broken downspout pouring onto a wall with a crack in it, or a faulty window sill) and vapor barriers.

As with any type of wall system, water infiltration is problematic. This can happen as the result of many and various occurrences: cracks in plaster skins, leaks in the roof above wall cavities, leaky plumbing, penetrations in plasters that attract moisture (gutters and poor plaster detailing are the number one culprits), poor window detailing, and poor flashing details. The best maintenance is prevention through quality work and good detailing. A small leak can grow into a much bigger problem if not detected, and serious leaks from plumbing, the roof, or other sources can cause large amounts of water to quickly concentrate in one spot in a wall. Any point load will cause rapid deterioration of a wall.

If you suspect a leak or notice a growing brown spot on your plaster, the first step would be to get a moisture meter and probe the wall to get an idea of where the moisture is coming from. From there, you can excavate the worst portion of the wall cavity to help dry it out. Moisture wants to reach an equilibrium: vapor will migrate from moist to dry until it finds a natural balance. You can help facilitate this with fans, dry heat, and perforations in the wall to help moisture escape. This can help bring the moisture level in the wall down and prevent any lasting damage.

If the water event has been continuous, and there is goopy black mold appearing in a portion of the wall, you'll need to completely remove any moldy LSC, and let the wall dry. Once dry (moisture content under 14% — use a moisture meter), apply industrial-strength hydrogen peroxide (3% solution). Let that dry, and, when conditions are right (summer, or when you have the ability to adequately climate control the area to be repaired) fill back in with a new mix of LSC.

Renovation

When the time comes to add a room, change a finish or remodel, all those photos you took of what is in the wall cavities will be worth their weight in gold! Remodeling and renovation of LSC structures is essentially the same as for any building in that you will likely encounter complexities due to unknowns within the walls and structures and the locations of wiring and pipes.

To make a new window or door opening: turn off all potential breakers to prevent a potential shock or fire situation when cutting through the wall. Depending on the thickness of the wall in question, you can use a blade specific to demolition (called a *demo blade*) on a reciprocating saw and cut through the wall just as you might a conventionally built wall. Antique bale saws also work. However, a long reciprocating saw blade has the tendency to follow the path of least resistance and might not make a perfectly straight plunge cut. A better route might be to mark out the opening and drop a circular saw with a sacrificial blade (i.e., an old blade you don't mind dulling from cutting though tough straw and soil) into the wall, cutting along the lines of the marked out opening. Then use a drill with a long bit (or use an extension bit) to drill through and transfer the corners of the cut to the opposite side of the wall, or transfer your measurements. Repeat with the circular saw on the opposite side and then use a reciprocating saw to cut the remaining light straw clay free and (with some help, as LSC is actually rather heavy, at about 30 pounds per cubic foot) push out the section. It might be easier to cut it out in two sections. Expect this to be a very dusty affair, and take appropriate precautions, both for the cleanliness of the room and for your own health and safety. The light straw clay that is removed can either be composted or rehydrated for reuse.

Since light straw clay is non-load bearing, all remodel or renovation to an existing project must connect with the preexisting framing, so this is very straightforward from a conventional building perspective.

If adding a window or a door opening, it is possible to do it very neatly and cleanly, without having to do a lot of plaster repairs. However, this can be a time to consider a new plaster color, clay, or lime paint to cover up any patchwork.

Window renovation.
PHOTO CREDIT: LYDIA DOLEMAN

Repairs

Common issues over time are dings and other damage to the protective plaster finishes — inside and out. Major repairs would likely be due to water damage.

Dings

In any building, lots of things happen that cause minor damage to walls: furniture gets moved around; pets and kids get rowdy; and many other small accidents occur. If the plaster skin is earthen, repair work is relatively easy. Ideally, you saved some "cookies" (small portions of the original plaster dried into small cookie-shaped patties), so you don't have to worry about color matching. Simply moisten the wall where the ding or chip has occurred by lightly misting the wall with a pump sprayer, garden hose on mist, or spray bottle. Only mist to the point where the wall surface is shiny, but not dripping. After three or more passes where you pause to let the water soak in, but not dry completely, the earthen plaster should be pliable, and you can take a small amount of the plaster cookie (also moistened — to a slightly runny cookie-dough consistency) and apply it to the wall (you can use a yogurt lid for a trowel). Let the section dry and lightly rub with a slightly damp sponge to blend in.

When you install type-S lime plaster walls ("hydrated" lime), it is a good idea to save a quart or two for repairs. You can use it following the directions just given for earthen plaster repairs. If the plaster is lime and there are no saved samples or the plaster is NHL (Natural Hydraulic Lime) lime plaster, you will have to make a new small batch of plaster. See Chapter 9, Finishes.

Sometimes a wall will *delaminate,* which is what it's called when a large portion of plaster comes off in a sheet. This is an indication of either frost damage (if on the exterior) or poor bonding between layers; sometimes it is an indication of some sort of structural shift. Delamination requires removal of any non-adhered plaster (you can soften it and reuse it, if earthen; if it is a type-S lime or NHL plaster, or if no samples exist, you will have to remake it). Once again, moisten the wall after determining that the issue isn't a substrate issue. (See "Holes, Shrinkage Gaps, and Soft Spots," just below for instructions if the layer of remaining plaster is spongy to the touch.)

Cracks can form due to a variety of causes: pulling too hard on the finish coat, substrate issues, freeze/thawing, incorrect ratios in the plaster or insufficient straw in the mix, or expansion and contraction in the building.

If the crack is singular and on the exterior of the building in an exposed area it, is a good idea to patch it quickly, as this crack can let a lot of water into the building and further deteriorate the plaster in that area.

Cracks routinely develop around window and door openings due to movement and uneven expansion and contraction around openings. Cracks can also form where two walls meet and they are expanding and contracting at different rates.

Holes, Shrinkage Gaps, and Soft Spots

Holes can occur due to too large amounts of light straw clay being crammed into a form at a time and not getting properly tamped into place. After all forms have been removed, a mix of slightly higher-slip light straw clay can be applied to the holes and smoothed over with a trowel. It is important to gently work the light straw clay into the hole so that it is fully integrated into the walls and won't shrink so much that as it dries it just falls out of the hole. This can be done by applying pressure when filling the hole, wiping the excess clay over the filled-in section, and spreading it into the surrounding LSC to help it to adhere.

Shrinkage gaps occur after the entire wall has dried and the clay has shrunk, causing small gaps to open between the light straw clay and the framing members or the horizontal reinforcing rods. This may indicate that too high-slip content was used, but sometimes that can't be avoided. Providing adequate key in the framing is important in limiting the shrinkage gaps. This is especially true of retrofits and projects where the light straw clay is being infilled between studs (as opposed to split studs or Larsen truss systems).

If the cracks are very fine but you can see light through them, make a scratch coat plaster mix of one part slip, one part fine sand, and one part chopped fiber; use this to patch over cracks.

Soft spots are the most challenging to repair. Hence, the preference for over-tamping rather than under-tamping in a temporary form system. Soft spots can also be an issue in projects incorporating permanent formwork. Soft spots will invariably lead to cracking and challenges with the plasters.

Worst-case scenario for repairing a soft spot is to pull out all the under-tamped light straw clay, place a large form on one side of the wall and start again using a heavy-slip mix to integrate between the new and the old light straw clay at the top and bottom of the repair section.

Best-case scenario: the soft spot is only slightly soft, and you can essentially firm it up with some of the repair mix that has a high fiber content. If the soft spot isn't "spotted" until plaster time, reinforcing the area with ¼″ mesh or some lath or something else very stable (especially if the stud bays are more than 24″ on-center layout) to firm up the substrate for plaster.

Chapter 11

Building Codes

THE PREMISE OF UNIVERSAL BUILDING CODES is that they create standards for human safety in buildings. Traditional technologies that preceded the advent of building codes are at an interesting crossroads where old world, time-tested, low-tech building techniques are being adapted to meet modern-day standards. Building with straw bales has been on the map, so to speak, a much shorter time than traditional LSC, but it is recognized by many more building departments. Part of that gap is due to a lack of information, testing, and industry support for LSC.

We live in an age when many sustainable living and building practices are not codified and are therefore "illegal," making it challenging if you are attempting to live and build by environmentally ethical standards. Thanks to the work of a handful of committed architects, engineers, and individuals, there are now legal precedents for LSC wall systems.

The state of New Mexico has had a LSC code for over a decade, and Portland, Oregon, approved LSC as a result of the work of the Alternative Technology Advisory Committee. However, by far the largest stride for LSC in the realm of building codes was the approval in 2013 of Appendix R — Light Straw-Clay Construction for the 2015 International Residential Code (IRC). (See Appendix 1 of this book for the full text of Appendix R)

The International Code Council (ICC) develops an array of model building codes, including the IRC, through public process. These codes are revised every three years. Though the IRC has no legal standing of its own, it is used as the basis for the residential code in virtually every jurisdiction in the United States. However, appendices in the IRC (like Appendix R) are not always adopted by state or local jurisdictions.

As this book goes to press (May 2017) at least four U.S. states have adopted Appendix R for statewide use — Maryland, New Jersey, New Mexico, and Oregon. Other states, such as California, have included it in their state residential code for optional adoption by local jurisdictions. Even if a jurisdiction has not adopted Appendix R, it can be proposed to a local building department for use on a project basis.

In October 2016 substantial revisions to Appendix R were approved by ICC for the 2018 IRC which will be published in September 2017. Its earliest adoption by a state, county or municipal jurisdiction is during the year 2018. However, the 2018 Appendix R can also be proposed for use on a project basis before jurisdictional adoption.

Regardless of whether your jurisdiction has adopted Appendix R, the building department might require a licensed design professional to take responsibility for your LSC design. If so, it is recommended you engage the services of an architect or structural engineer who is familiar with LSC

Whenever an LSC project is proposed where Appendix R has not been adopted, it is important to understand your local building codes and to engage early on with your building official. If your building official is not familiar with LSC, it is equally important to provide information to answer his or her questions and demonstrate

LSC as a safe building material and system. This includes, but is not limited to sharing Appendix R, and especially its commentary, which explains its requirements. Testing reports and books on LSC can also be very valuable for this purpose.

In places where building codes are not required (sometimes very rural jurisdictions), builders are much freer to build as they choose. However, even in these cases, Appendix R is a good reference for building safely with LSC, as is this book and others listed on its Resources page.

Chapter 12

Tools

THE FOLLOWING ARE LISTS of all required tools — from basic to industrial (not including tools for framing). Having the right tools for the job makes the project flow smoothly. As is the case with most unique building systems, some tools are manufactured on the job, as necessity is the mother of invention. Most of the tools needed for installing light straw clay are easy to acquire and simple; some, like the tumbler, require a skilled person to construct.

Tools to Process Clay

For a basic 100-square foot shed:

- ¼"–½" screen
- spades and flat shovels
- at least five 5-gallon buckets
- one 30-gallon garbage can, or one 55-gallon drum with an open top
- mixing drill or wheelbarrow and hoe
- access to water for mixing and cleanup

For project implementing a tumbler:

- ¼" to ½" screen
- spades and flat shovels
- at least five 5-gallon buckets
- one 55-gallon drum with open top
- mixing drill
- pond pump and electricity to power it (some are gas)
- scaffolding
- 100–300 gallon trough
- 15–20′ of 4″ tubing for moving slip to trough
- 2× supports for drain tube
- access to water for mixing and cleanup

Tools to Process Straw

For a basic 100-square foot shed:

- pallets, or 2× material for keeping straw off the ground
- sufficient tarps and or equivalent for keeping straw dry until use
- baling hooks and gloves for moving bales
- utility knife for cutting twine
- dust masks

For bigger projects implementing a tumbler:

- possible rental of a squeeze for more than 100-bale projects
- tractor or truck for moving bales
- pallets, or 2× material for keeping straw off the ground
- sufficient tarps and/or equivalent for keeping straw dry until use
- baling hooks and gloves for moving bales
- utility knife for cutting twine
- dust masks

Tools for Mixing Light Straw Clay

For a basic 100-square foot shed:

- 5-gallon bucket for transporting slip from slip mixing area to table
- saw horses
- sheet of plywood
- tarp to cover plywood
- dust masks
- gloves
- tarp or receptacle for mixed light straw clay
- water for cleanup

For bigger projects implementing a tumbler:

- pond/ditch pump
- 4″ tubing long enough to transport slip to holding tank
- electricity or gas
- a tumbler (See Resources for information on how to make a tumbler; you can sometimes find one to rent if you ask around.)
- plywood platform for landing, or custom-built box to load a tractor
- tractor or tarp or wheelbarrow for transport
- shade for tumbler operator
- water for cleanup
- scaffolding to support holding tank
- 100–300 gallon tub/trough for holding slip
- utility knives for cutting open straw bales
- dust masks

Tools for Temporary Forms

For a basic 100-square foot shed:

- enough sheets of ⅝″ plywood to wrap the bottom perimeter of the project inside and out (perimeter of building × 2, divided by 8 = # of sheets of plywood needed)
- # of sheets of plywood × 2 = number of 8′ 2×4s to make "strongbacks" for forms
- circular saw to rip forms down to height
- drill or impact driver for fastening 2×4 to plywood to make strongbacks
- Forty 2¼″ hex head washered screws per sheet of plywood
- hex head driver for fasteners (good to have extra and paint them a bright color, so you can find them if they get dropped!)
- impact driver for fastening forms to studs or vertical members
- bucket of water and towel to keep hands clean and dry
- ear and eye protection
- ladders or scaffolding for tall wall sections
- basic carpentry tools for specialty or problem areas (jig saw for electrical or plumbing cut outs in formwork; pry bar for broken, stripped, or stuck screws; hammer for nails sticking out of studs)

For bigger projects implementing a tumbler:
Same as above, but you'll need more than one fastening impact driver for the formwork, and more ladders and scaffolding.

Tools for Permanent Forms

For a basic 100-square foot shed with reed mat:

- reed matt sufficient to cover both sides of the project (add 10% for openings and gable ends)
- a pneumatic stapler with a 1″ or 7⁄16″ crown
- box of 10,000 galvanized staples (1″ or longer)
- 800 linear feet of medium gauge wire or a spool of polypropylene twine
- compressor
- hose for compressor and nail gun
- pneumatic gun oil
- ear and eye protection
- scissors/snips
- utility knife
- sheet of plywood and a 7″ wide strip of plywood
- tape measure
- masking tape
- marker

For a basic 100-square foot shed with wood lath:

- Wood lath (wood or plywood ¼″–⅝″ thick by 1½″ by however long you can get it)
- a pneumatic stapler with a 1″ or 7⁄16″ crown
- box of 10,000 galvanized staples (1″ or longer)
- compressor

- hose for compressor and gun
- pneumatic gun oil
- ear and eye protection
- circular saw or chop saw for cutting lath to length
- measuring tape
- pencil

For bigger projects implementing a tumbler:
The same as above for both reed mat and wood lath.

Basic Tools for Tamping

- 2×4 scrap, from 8″ long to 36″ long

Basic Tools for Cleanup

- hose, sponges, and scrubbers for small projects
- buckets for soaking tools
- a pressure washer for bigger projects (Beware the jet of water from a pressure washer! It can mar wood and cut through tarps and skin.)

Moisture Monitoring

The Delmhorst bale probe can be used to get an idea of the moisture content in a LSC wall. It isn't technically calibrated for LSC, but during the drying phase can be used to assess if moisture levels are dropping.

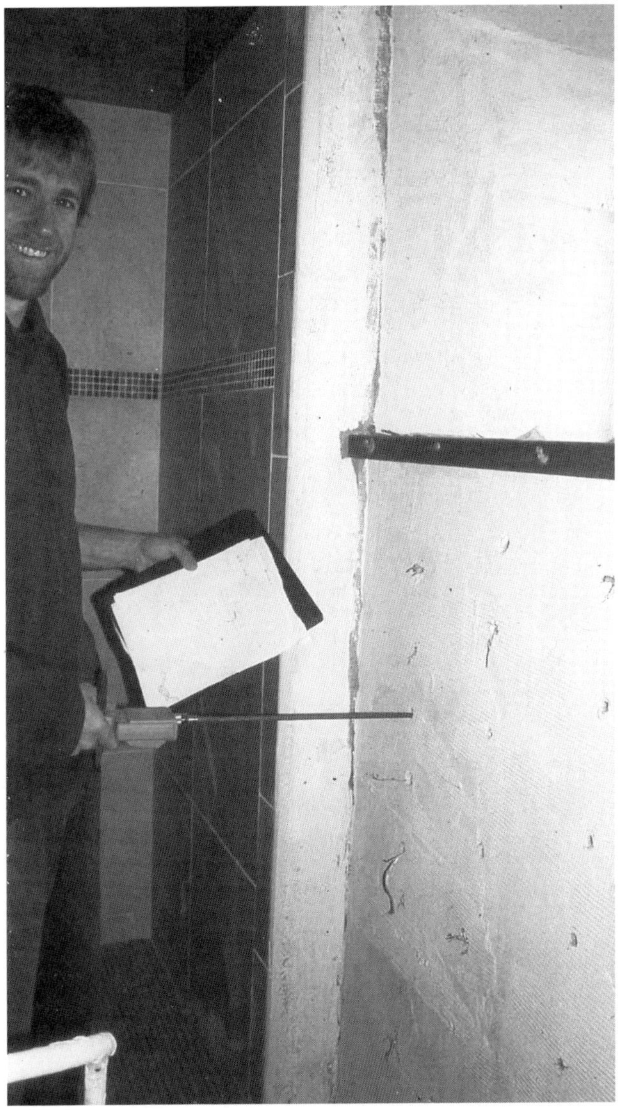

Using a moisture meter. Photo credit: Lydia Doleman

Appendix 1: From the 2015 IRC with Commentary

Appendix R — Light Straw-Clay Construction (also see Chapter 11; pages 95-6)

(The provisions contained in this appendix are not mandatory unless specifically referenced in the adopting ordinance.)

General Comments

Mixtures of clay and straw known as clay-straw, straw-clay, and light straw-clay have been successfully used as a nonload-bearing wall infill material since 1950 in Europe and 1990 in the United States. The use of light straw-clay wall systems has seen a surge in popularity recently because of its attributes. These include light straw-clay's thermal performance (with its balance of mass and insulation) and its low environmental impact, using local, minimally processed materials that ultimately return to the earth without harm. Other beneficial qualities include the low cost of the raw materials and the easily-learned skills needed for installing the material.

Two heavier forms of wall construction using clay and straw have been used in various parts of the world for thousands of years. A monolithic clay and straw wall system known as cob has been used in what are now the United Kingdom, Africa, the Middle East and eastern North America. A system of woven wood or woven bamboo with applied clay and straw known as wattle and daub was in common use for centuries throughout Europe, Africa, Asia and North and South America. Both cob and wattle and daub continue to be utilized today. Many thousands of existing structures using cob or wattle and daub, dating back 300-400 years, have been continuously occupied, attesting to the durability of the materials of clay and straw.

Purpose

In the United States, residential and nonresidential structures using light straw-clay have been completed in 17 states, and most of those have been constructed with full permits and inspections. However, many of these structures have been constructed on a case-by-case basis without t he benefit of a building code for light-straw clay construction.

As of 2014, only the states of New Mexico and Oregon had adopted guidelines or codes for light straw-clay construction. This appendix provides prescriptive and performance requirements for all aspects of non-bearing light straw clay construction. This is especially important because most design professionals, code officials and builders are currently unfamiliar with this wall system.

The purpose of this appendix and its commentary is to provide requirements and guidance regarding construction utilizing straw and unfired clay, which are materials previously not addressed in the code. The nature of light straw-clay as a material and system demands compliance with certain requirements that differ from established practices with conventional materials and systems in the code. The commentary to this appendix is particularly useful in these instances because it explains the reasons for requirements that are new to users of the code and this appendix.

SECTION AR101
GENERAL

AR101.1 Scope. This appendix shall govern the use of light straw-clay as a nonbearing building material and wall infill system in Seismic Design Categories A and B.

❖ In Seismic Design Categories A and B, lateral loads due to wind govern the requirements for wall bracing. In Seismic Design Categories C, D_0, D_1 and D_2 seismic loading often governs. A building's performance when subjected to lateral wind forces is independent of the building's weight, whereas a building's performance when subjected to seismic forces is a function of the building's weight. A wood frame wall with light straw-clay infill is significantly heavier than a conventional wood frame wall for which the code's bracing requirements are formulated. Thus, this appendix is limited to use in Seismic Design Categories A and B, where the relative increased weight of light straw-clay walls is not relevant, and the bracing requirements in the code text and tables remain applicable.

Buildings using light straw-clay construction in Seismic Design Categories C, D_0, D_1 and D_2 are allowed if the building official finds that the proposed design satisfies the requirements of Section R104.11, Alternative materials, design and methods of construction and equipment. This would likely require an approved engineered design for wall bracing that accounts for the weight of light straw-clay walls. However, because this appendix is limited to Seismic Design Categories A and B for wall bracing requirements only, all other requirements of this appendix, where adopted, would apply to a building using light straw-clay construction in Seismic Design Categories C, D_0, D_1 and D_2.

In general, many variations of light straw-clay construction are possible and many have been practiced historically. Designs that do not comply with certain requirements in this appendix may be approved if the building official finds the proposed design satisfies Section R104.11 by means of equivalent alternatives to these requirements.

SECTION AR102 DEFINITIONS

AR102.1 General. The following words and terms shall, for the purposes of this appendix, have the meanings shown herein. Refer to Chapter 2 of the *International Residential Code* for general definitions.

CLAY. Inorganic soil with particle sizes of less than 0.00008 inch (0.002 mm) having the characteristics of high to very high dry strength and medium to high plasticity.

CLAY SLIP. A suspension of clay soil in water.

CLAY SOIL. Inorganic soil containing 50 percent or more clay by volume.

❖ The word "soil" is commonly associated with topsoil, which contains organic matter. Clay soil used for construction purposes is a non-fertile (inorganic) mineral subsoil that contains clay, silt and sand. Sections AR103.3.2, and AR103.4.1 and Commentary Figure AR103.3.2 describe methods for determining the suitability of this material for light straw-clay construction.

INFILL. Light straw-clay that is placed between the structural members of a building.

LIGHT STRAW-CLAY. A mixture of straw and clay compacted to form insulation and plaster substrate between or around structural and nonstructural members in a wall.

❖ The density range of light straw-clay is 10 to 50 pounds per cubic foot (160 to 801 kg/m^3).

NONBEARING. Not bearing the weight of the building other than the weight of the light straw-clay itself and its finish.

STRAW. The dry stems of cereal grains after the seed heads have been removed.

VOID. Any space in a light straw-clay wall in which a 2-inch (51 mm) sphere can be inserted.

SECTION AR103 NONBEARING LIGHT STRAW-CLAY CONSTRUCTION

AR103.1 General. Light straw-clay shall be limited to infill between or around structural and nonstructural wall framing members.

AR103.2 Structure. The structure of buildings using light straw-clay shall be in accordance with the *International Residential Code* or shall be in accordance with an *approved* design by a registered *design professional*.

AR103.2.1 Number of stories. Use of light straw-clay infill shall be limited to buildings that are not more than one *story above grade plane*.

> **Exception:** Buildings using light straw-clay infill that are greater than one *story above grade plane* shall be in accordance with an approved design by a registered *design professional*.

AR103.2.2 Bracing. Wind shall be in accordance with Section R602.10 and shall use Method LIB. Walls with light straw-clay infill shall not be sheathed with solid sheathing.

❖ The "solid sheathing" prohibition in this section refers only to the application of sheathing that will inhibit air-drying of newly installed straw-clay material. Because light straw-clay is installed wet and is not fully dried to equilibrium (see AR103.5.1) before the application of plaster, any solid impermeable permanent sheathing that would inhibit its drying is prohibited.

AR103.2.3 Weight of light straw-clay. Light straw-clay shall be deemed to have a design dead load of 40 pounds per cubic foot (640 kg per cubic meter) unless otherwise demonstrated to the *building official*.

❖ Densities of light straw-clay can vary depending on wall design, project location and desired thermal properties. Historically, densities ranging from 10 to 50 pounds per cubic foot (160 to 801 kg/m^3) have been utilized. Structural engineering is not required for one-story light straw-clay structures that comply with the prescriptive structural requirements of the code and this appendix. However, structural calculations may be needed when departing from these requirements, or may be otherwise advantageous.

Densities of light straw-clay stated in this appendix are for the straw-clay material only and do not include other elements such as the wall's plaster or other finish and its framing.

AR103.2.4 Reinforcement of light straw-clay. Light straw-clay shall be reinforced as follows:

1. Vertical reinforcing shall be not less than nominal 2-inch by 6-inch (51 mm by 152 mm) wood members at not more than 32 inches (813 mm) on center where the vertical reinforcing is nonload bearing and at 24 inches (610 mm) on center where it is load bearing. The vertical reinforcing shall not exceed an unrestrained height of 10 feet (3048 mm) and shall be attached at top and bottom in accordance with Chapter 6 of the this code. In lieu of these requirements, vertical reinforcing shall be in accordance with an *approved* design by a registered *design professional*.
2. Horizontal reinforcing shall be installed in the center of the wall at not more than 24 inches (610 mm) on center and shall be secured to vertical members. Horizontal reinforcing shall be of any of the following: $^3/_4$-inch (19.1 mm) bamboo, $^1/_2$-inch (12.7 mm) fiberglass rod, 1-inch (25 mm) wood dowel or nominal 1-inch by 2-inch (25 mm by 51 mm) wood.

❖ The term "reinforcement" in this section means a component of the wall system that contains or keeps the light straw-clay infill in place during construction and in service. This includes containment of the light straw-clay where the wall is subjected to lateral out-of-plane wind or seismic forces as required by the code.

Item 1: Vertical reinforcing can be composed of nonload-bearing elements, with containment of the light straw-clay as their sole function, or elements that also serve a structural function, such as load-bearing studs. Vertical reinforcing and wall frame construction are based on the conventional wood framing requirements of Section R602, with similar methods and requirements for studs, plates and headers, etc. The main departure from the requirements in Section R602 in light straw-clay construction is the constraint on the use of solid sheathing, as noted in the commentary to Section AR103.2.2. This section is not intended to prohibit commonly accepted vertical reinforcing methods that are structurally equivalent to 2-inch by 6-inch solid wood members, such as "Larsen trusses" or similar composite members. Commentary Figures AR103.2.4(1), AR103.2.4(2) and AR103.2.4(3) show examples of wall framing for light straw-clay construction.

Item 2: Unlike the placement of steel reinforcing bar in typical concrete construction, which is used for tensile and shear reinforcing, horizontal reinforcing members in light straw-clay are used primarily for the doweling effects of the reinforcing within the wall system. The horizontal reinforcing helps: (1) Maintain overall integrity of the wall assembly during construction and in service; (2) Transfer of out-of-plane wind or seismic forces from the light straw-clay mass to the vertical reinforcing framing members and; (3) Control the vertical shrinkage and settling of the light straw-clay material as it dries. In practice, the horizontal reinforcing is placed while slightly compressing the moist straw-clay mix in order to limit "rebound" and help control shrinkage.

AR103.3 Materials. The materials used in light straw-clay construction shall be in accordance with Sections AR103.3.1 through AR103.3.4.

AR103.3.1 Straw. Straw shall be wheat, rye, oats, rice or barley, and shall be free of visible decay and insects.

❖ Straw, like wood, is composed primarily of cellulose and lignin. Construction-grade straw can be readily identified by visual examination. The best straw is composed of the dry (i.e. not having been wetted since cut) intact stems only of wheat, rye, oats, rice or barley plants and must be free of visible decay and contaminants such as insects, fertile topsoil and green plant material, since mold and mildew can grow in the presence of these readily-digested food sources.

Straw is best harvested dry and kept dry until it is mixed with clay slip and installed. The straw will have a golden color and a glossy surface. Excessive exposure to moisture will result in straw with reduced color and gloss. This straw is not optimal for light straw-clay construction.

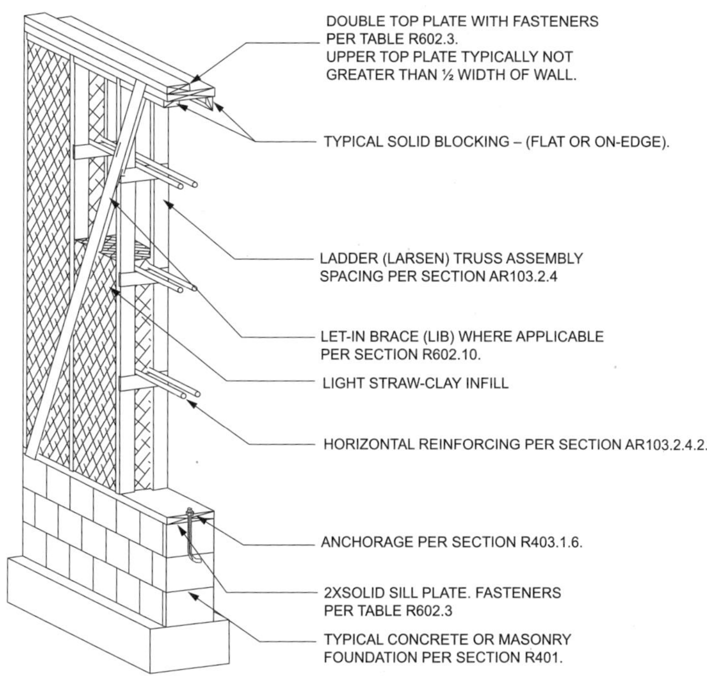

Figure AR103.2.4(1)
LIGHT-STRAW-CLAY WALL WITH LARSEN TRUSS

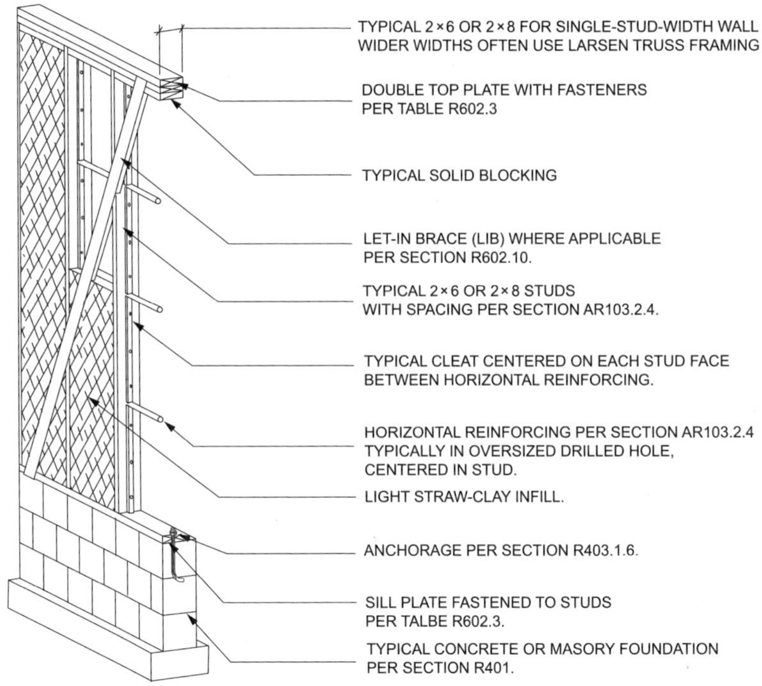

Figure AR103.2.4(2)
LIGHT STRAW-CLAY WALL WITH SINGLE-STUD WIDTH

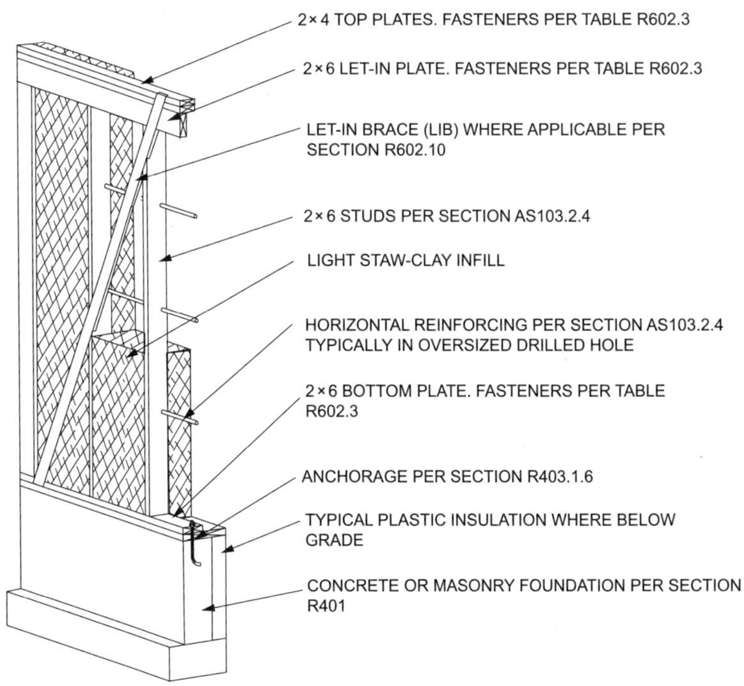

Figure AR103.2.4(3)
LIGHT STRAW-CLAY WALL WITH BLIND STUD

AR103.3.2 Clay soil. Suitability of clay soil shall be determined in accordance with the Figure 2 Ribbon Test or the Figure 3 Ball Test of the Appendix to ASTM E2392/E2392M.

❖ This section is intended to ensure that the quality of the clay slip used to make light straw-clay is suitable for the intended application. Clay soil appropriate for light straw-clay construction contains particles of sand and silt in addition to microscopic particles of clay. However, only the microscopic clay particles that form the protective bonds around and between the individual strands of straw, ensure stability and longevity. The percentage of clay in the clay slip will determine the effectiveness of the coating and binding of the straw. See Commentary Figure AR103.3.2 for recommended mixture ratios and testing methods.

A ribbon or ball test, as stated in this section, is sufficient to determine suitability for wall densities between 30 and 50 pounds per cubic foot (480 to 801 kg/m^3). A lab test known as a "soil texture test" or "physical analysis test" yields an even more accurate analysis, giving precise percentages of clay, sand and silt.

A lab test that measures percentages of sand, silt and clay is strongly recommended for lower-density, higher-insulating light straw-clay mixes. This is because there is much less clay soil added to the mix, so the little clay soil present must have a relatively high percentage of clay to provide the protective bonds that ensure stability and longevity of the wall. Inexpensive soil texture or physical analysis tests are readily available through university cooperative extension and agricultural soil laboratories in most parts of the U.S. and Canada.

AR103.3.3 Clay slip. Clay slip shall be of sufficient viscosity such that a finger dipped in the slip and withdrawn remains coated with an opaque coating.

❖ This requirement is intended to provide a simple field method to verify that enough clay is present in the slip to achieve adequate coverage of the straw and ensure cohesive integrity of the installed light straw-clay mix. This test works for all densities of light straw-clay.

AR103.3.4 Light straw-clay mixture. Light straw-clay shall contain not less than 65 percent and not more than 85 percent straw, by volume of bale-compacted straw to clay soil. Loose straw shall be mixed and coated with clay slip such that there is not more than 5 percent uncoated straw.

❖ The intent of this section is to describe straw and clay percentages for densities in the 30 to 50 pounds per cubic foot range (480 to 801 kg/m^3). The percentage of straw described in this section means straw-plus-air in the wall. These percentages are not meant to limit higher percentages of straw-plus-air used in lighter mixes where the subsoil content in the wall is lower and meets the recommended testing in Commentary Figure AR103.3.2. Commentary Figure AR103.3.4 is useful for clarifying the percentages of components used to achieve the range of light straw-clay densities.

AR103.4 Wall construction. Light straw-clay wall construction shall be in accordance with the requirements of Sections AR103.4.1 through AR103.4.7.

AR103.4.1 Light straw-clay maximum thickness. Light straw-clay shall be not more than 12 inches (305 mm) thick, to allow adequate drying of the installed material.

❖ Historically, 12 inches (305 mm) has been considered the maximum wall thickness for straw-clay mixtures 30 pcf and greater. Current research suggests that higher percentages of clay in the slip allows for lighter density

DENSITY (pcf)	MINIMUM % CLAY IN SUBSOIL (1)	MINIMUM CLAY: SILT RATIO	CLAY TESTING METHOD (2)	MAXIMUM LIGHT STRAW-CLAY (thickness, inches)
10	70	3.5:1	A	15
12	46	1.7:1	A	15
13	40	1.33:.1	A	15
15	35	0.95:1	A	15
20	30	0.60:1	A	12
30	25	0.42:1	A or B	12
40	Test Method B	Test Method B	B	12
50	Test Method B	Test Method B	B	12

Courtesy of Douglas Piltingsrud@StrawClay.org. Used by Permission.

1. The minimum percentage criteria may be met by using unprocessed ray clay subsoil or subsoil refined by removal of sand and silt.
2. Subsoil Testing Methods:
   a. Lab test for percent of clay, silt and sand via hydrometer method.
   b. Ribbon test or the Figure 3 Ball Test in the Appendix of ASTM E2392/E2392M.

Figure AR103.3.2
RATIOS AND TESTING OF LIGHT STRAW-CLAY MIXTURES

straw-clay mixtures (with higher *R*-values) while still achieving adequate drying in walls up to 15 inches (381 mm) thick. Section AR103.4.1 is intended to ensure safe drying parameters for denser mixes (30 pcf and greater), and is not intended to limit the wall thickness for lighter mixtures where a history of successful drying of walls with lower densities that exceed 12-inches (305 mm) is demonstrated to the building official.

See Commentary Figure AR103.3.2 for the recommended maximum thicknesses for light straw-clay of various densities. This figure also shows the relationships between light straw-clay components and the recommended clay soil testing method for those densities.

AR103.4.2 Distance above grade. Light straw-clay and its exterior finish shall be not less than 8 inches (203 mm) above exterior finished *grade*.

AR103.4.3 Moisture barrier. An *approved* moisture barrier shall separate the bottom of light straw-clay walls from any masonry or concrete foundation or slab that directly supports the walls. Penetrations and joints in the barrier shall be sealed with an *approved* sealant.

❖ Moisture barriers to protect the straw-clay from the rising dampness inherent to masonry or concrete foundations include materials historically used for this purpose such as asphalt-impregnated roofing felt, plastic sheeting, elastomeric gasketing, clay, tar, and liquid applied products manufactured for use as a moisture barrier over concrete or masonry.

AR103.4.4 Contact with wood members. Light straw-clay shall be permitted to be in contact with untreated wood members.

❖ Light straw-clay installation is similar to the installation of wet-blown cellulose insulation in that both are installed damp, dry in the following weeks and then remain dry throughout their service life. In addition, clay has a long history of preserving wood. Clay is hydrophilic (i.e., water-attracting) and is highly water-blocking due to its strong surface bonding with any available water. Also, because of clay's very large total surface area (due to its layered microscopic structure), it has considerable capacity to capture and store water. Even when straw-clay is applied wet and in full contact with wood, the clay functions to pull water away from adjacent materials as the composite dries, including wood and its own straw fibers. In this way, and with its ability to act as a water barrier (e.g., clay is used to line ponds to prevent leakage), the clay keeps the moisture in adjacent materials below the levels necessary to support microbe growth (i.e., rot) and, therefore, acts as a preservative (see commentary, Section AR103.5.3).

AR103.4.5 Contact with nonwood structural members. Nonwood structural members in contact with light straw-clay shall be resistant to corrosion or shall be coated to prevent corrosion with an *approved* coating.

❖ A light straw-clay installation is similar to a wet-blown cellulose installation in that the material is damp only temporarily. This limited wetting is a common construction condition and standard coatings such as primer paint and electro-galvanizing are acceptable to prevent corrosion of nonwood structural members. In this section, the term "nonwood" is intended to mean metal that is susceptible to corrosion with prolonged exposure to moisture.

AR103.4.6 Installation. Light straw-clay shall be installed in accordance with the following:

1. Formwork shall be sufficiently strong to resist bowing where the light straw-clay is compacted into the forms.
2. Light straw-clay shall be uniformly placed into forms and evenly tamped to achieve stable walls free of voids. Light straw-clay shall be placed in lifts of not more

LIGHT STRAW-CLAY DENSITY (pcf)	STRAW (pcf)	SUBSOIL (pcf)	SUBSOIL (1) VOLUME %	STRAW (2) VOLUME %	AIR VOLUME %	STRAW + AIR VOLUME %	*R*-VALUE (hr/F°/cu.ft./BTU/inch)
10	6.7	3.3	2.0	7.4	90.6	98.0	1.80
12	6.7	5.3	3.1	7.4	89.4	96.9	1.72
13	6.7	6.3	3.7	7.4	88.9	96.3	1.69
15	6.7	8.3	4.9	7.4	87.7	95.1	1.63
20	6.7	13.3	7.9	7.4	84.7	92.1	per AR104.1
30	6.7	23.3	13.8	7.4	78.7	86.2	per AR104.1
40	6.7	33.3	19.9	7.4	72.8	80.2	per AR104.1
50	6.7	43.3	25.7	7.4	66.9	74.3	per AR104.1

Courtesy of Douglas Piltingsrud@StrawClay.org. Used by Permission.

1. Uses 168 pcf for subsoil specific (2.7 g/cc).
2. Uses 90 pcf for straw gravity (1.45 g/cc). Straw Volume % and associated columns may increase for 20 pcf walls and above due to the weight from additional subsoil.

Figure AR103.3.4
CHARACTERISTICS OF LIGHT STRAW-CLAY AT VARYING DENSITIES

than 6 inches (152 mm) and shall be thoroughly tamped before additional material is added.

3. Formwork shall be removed from walls within 24 hours after tamping, and walls shall remain exposed until moisture content is in accordance with Section AR103.5.1. Visible voids shall be patched with light straw-clay prior to plastering.

❖ **Item 1:** Commonly used formwork for light straw-clay installations includes plywood, OSB panels and solid wood boards; often in practice these materials are then repurposed as subflooring and roof sheathing. In contrast to poured concrete, which is liquid and extremely heavy, light straw-clay is semi-solid, lightweight and is placed using only a person's body weight to tamp it into place. Accordingly, the formwork for light straw-clay is typically constructed much more lightweight than concrete formwork. For example, it often uses only 2 ☒ 4s for whalers and is attached temporarily with screws to the permanent wall framing as part of the formwork's structure and bracing.

Typically, formwork is first installed completely on one side of the wall, then incrementally on the other side from the bottom and then upward as the straw-clay is also incrementally placed. This allows a person to tamp the material standing partially inside the formwork, and allows for possible reuse of the lower forms ("slip forming") once the material in the form above is tamped in place (see commentary, Section AR103.4.6, Items 2 and 3).

Item 2: This item is intended to ensure the mix is compacted using light tamping only; to achieve the desired density, consolidate the mix, ensure adhesion between stalks of straw and achieve wall integrity without crushing the straw stalks. Intact straw stalks and the air spaces inside and between them ensure that the maximum wall *R*-value is achieved for that density of mix. For this reason, compaction is not normally accomplished by mechanical means. Rather, it is done by a person standing inside the formwork and "walking in the forms" as the straw-clay mix is inserted. The installer ensures complete distribution of material to all corners, making sure that no voids are left, and uses his or her body weight to lightly compress the straw-clay to the desired density.

Item 3: This section is intended to allow the light straw-clay to begin drying as soon as possible after placement and to avoid the possibility of mold growth on the surface of the straw-clay beneath the forms. Forms can generally be removed as soon as the straw has "set" (i.e., moist stalks have relaxed into place). Optimum time is generally 1 to 4 hours after placement.

AR103.4.7 Openings in walls. Openings in walls shall be in accordance with the following:

1. Rough framing for doors and windows shall be fastened to structural members in accordance with the *International Residential Code*. Windows and doors shall be flashed in accordance with the *International Residential Code*.
2. An *approved* moisture barrier shall be installed at window sills in light straw-clay walls prior to installation of windows.

AR103.5 Wall finishes. The interior and exterior surfaces of light straw-clay walls shall be protected with a finish in accordance with Sections AR103.5.1 through AR103.5.5.

AR103.5.1 Moisture content of light straw-clay prior to application of finish. Light straw-clay walls shall be dry to a moisture content of not more than 20 percent at a depth of 4 inches (102 mm), as measured from each side of the wall, prior to the application of finish on either side of the wall. Moisture content shall be measured with a moisture meter equipped with a probe that is designed for use with baled straw or hay.

❖ This provision is intended to require sufficient drying of the straw-clay in order to achieve dimensional stability and a stable surface for a noncracking plaster application. However, a "full-dry" condition (in full equilibrium with ambient conditions) is not needed for successful plastering, because plasters as required in Section AR103.5.2 are highly permeable and allow the straw-clay to continue drying after plastering.

It is not always possible to obtain consistent moisture meter readings if only a few locations are sampled. Sampling multiple locations yields more reliable information about the overall dryness, particularly as the less-dense mixes offer less surface contact area for the probes of electrical moisture meters. As an additional measure, vertical wall shrinkage should be observed as the material dries and plastering should begin only when shrinkage has stopped and any voids are filled. Depending on wind, temperature and humidity conditions, the time required for straw-clay to reach a plaster-ready state may vary from a few weeks to several months.

AR103.5.2 Plaster finish. Exterior plaster finishes shall be clay plaster or lime plaster. Interior plaster finishes shall be clay plaster, lime plaster or gypsum plaster. Plasters shall be permitted to be applied directly to the surface of the light straw-clay walls without reinforcement, except that the juncture of dissimilar substrates shall be in accordance with Section AR103.5.4. Plasters shall have a thickness of not less than $^1/_2$ inch (12.7 mm) and not more than 1 inch (25 mm) and shall be installed in not less than two coats. Exterior clay plaster shall be finished with a lime-based or silicate-mineral coating.

❖ The purpose of a lime-based or silicate-mineral coating over exterior clay plaster is to improve the plaster's resistance to erosion from weather. This section does not require exterior plaster, but rather describes its characteristics if used.

Some admixtures in clay plasters have a history of imparting desirable plaster characteristics, such as enhanced moisture resistance, durability and crack resistance. These admixtures include wheat paste, manure, casein and prickly pear juice. It is not the intention of this section to exclude the use of these or other admixtures that can demonstrate erosion resistance from weather equivalent to lime-based or silicate-mineral coatings.

The centuries-old European predecessors and the light straw-clay buildings built to date in North America have all been constructed without the use of a water-resistive barrier. The light straw-clay materials in this appendix are highly vapor permeable, which allows passage of water vapor without harm to the wall. Code precedents exist for construction systems such as adobe construction and log construction, which do not require the use of a water-resistive barrier. In these systems, as in light straw-clay construction, there is sufficient moisture buffer capacity to hold and re-release moisture without damage to structural members or degradation of the wall due to weather-related moisture. For these reasons, a water-resistive barrier is not required in light straw-clay construction.

AR103.5.3 Separation of wood and plaster. Where wood framing occurs in light straw-clay walls, such wood surfaces shall be separated from exterior plaster with No.15 asphalt felt, Grade D paper or other approved material except where the wood is preservative treated or naturally durable.

Exception: Exterior clay plasters shall not be required to be separated from wood.

❖ Clay is hydrophilic (i.e., water-attracting, drawing moisture to it and away from adjacent materials) and is highly water-blocking due to its strong surface bonding with any available water. In addition, because of clay's very large total surface area (due to its layered microscopic structure), it has considerable capacity to capture and store water. Historical examples include the centuries-old half-timbered buildings still in use in Germany, with straw-clay infill and clay plasters in direct contact with the original wood frame structure. This combination of hydrophilic properties, water-blocking properties and water storage capacity protects adjacent materials that might otherwise be harmed by extended exposure to moisture. Because of this, clay plasters are not required to be separated from wood (see also commentary, Section AR103.4.4).

AR103.5.4 Bridging across dissimilar substrates. Bridging shall be installed across dissimilar substrates prior to the application of plaster. Acceptable bridging materials include: expanded metal lath, woven wire mesh, welded wire mesh, fiberglass mesh, reed matting or burlap. Bridging shall extend not less than 4 inches (102 mm), on both sides of the juncture.

❖ This section is intended to prevent cracking of the plaster where it spans across wall framing members or substrates that interrupt the face of the installed light straw-clay. The bridging materials provide greater tensile and shear strength for plasters in areas that are prone to differential movement and provide a material to which the plasters can mechanically bond across substrates that otherwise offer little or no means for bonding.

AR103.5.5 Exterior siding. Exterior wood, metal or composite material siding shall be spaced not less than $^{3}/_{4}$ inch (19.1 mm) from the light straw-clay such that a ventilation space is created to allow for moisture diffusion. The siding shall be fastened to wood furring strips in accordance with the manufacturer's instructions. Furring strips shall be spaced not more than 32 inches (813 mm) on center, and shall be securely fastened to the vertical wall reinforcing or structural framing. Insect screening shall be provided at the top and bottom of the ventilation space. An air barrier consisting of not more than $^{3}/_{8}$-inch-thick (9.5 mm) clay plaster or lime plaster shall be applied to the light straw-clay prior to the application of siding.

❖ Plaster installed in a continuous layer over the interior or exterior light straw-clay surface meets the intent of this section, which is to provide the required air barrier.

SECTION AR104 THERMAL INSULATION

AR104.1 R-value. Light straw-clay, where installed in accordance with this appendix, shall be deemed to have an *R*-value of 1.6 per inch.

❖ Current research indicates that *R*-value in light straw-clay varies predictably with density. Higher percentages of clay in the slip and improved quality control of the light straw-clay mix allows for lighter density straw-clay mixtures with higher *R*-values. These lighter mixtures are particularly well suited to colder climate regions. This appendix does not intend to limit the use of these lighter and more insulating mixtures. Commentary Figure AR103.3.4 summarizes the *R*-value findings for densities from 10 to 15 pounds per cubic foot from the 2004 USDA Forest Products Laboratory *Engineering Report of Light Clay Specimens* (see the bibliography) and may be used for calculating thermal performance for walls using these lighter density mixes. The *R*-value stated in AR104.1 and those in Commentary Figure AR103.3.4 are for the light straw-clay material only, and exclude any additional thermal resistance provided by plaster or other finish over the light straw-clay.

SECTION AR105 REFERENCED STANDARD

ASTM E2392/
E2392M—10 Standard Guide for Design of Earthen Wall Building Systems AR103.3.2

Bibliography

The following resource materials were used in the preparation of the commentary for this appendix:

Baker-Laporte, Paula; Laporte, Robert. *Econest, Creating Sustainable Sanctuaries of Clay, Straw and Timber*. Layton, UT: Gibbs Smith, 2005.

APPENDIX R

Duncan, Richard, PE. *Resistance to Out-Of-Plane-Lateral Forces of Light Straw Clay Wall Infill.* http://www.econesthomes.com/wp-content/uploads/2013/01/Light-Straw-Clay-Out-of-Plane-Study-FINAL_merged.pdf, 2013.

Host-Jablonski, L. *Typical Outline Specifications for a Northern Light Straw-Clay House*. In *The Affordable Natural House Contractor Training Reference Manual,* 2nd Edition. S. Thering ed. Madison WI: University of Wisconsin-Extension, 2008.

Piltingsrud, D. and L. Host-Jablonski, *An Introduction to the Science of Northern Light Straw-Clay Construction*. In *The Affordable Natural House Contractor Training Reference Manual,* 2nd Edition. S. Thering, ed. Madison, WI: University of Wisconsin-Extension. 2008.

State of New Mexico Construction Industries Division Clay Straw Guidelines. Santa Fe, NM: State of NM CID publication. 2001.

StrawClay.org website. Madison, Wisconsin: Design Coalition, Inc., 2014.

Thornton, J., *Initial Material Characterization of Straw Light Clay*. Ottawa, Ontario: Canada Mortgage and Housing Corporation, 2004.

USDA Forest Products Laboratory *Engineering Report of Light Clay Specimens: Thermal Conductivities for Design Coalition's Straw/Clay Formulations Extend Volhard's K-Value vs. Density Curve in Low Conductivity End.* Madison WI: USDA Forest Products Laboratory. 2004.

2011 Oregon Reach Code—Section 1307. Country Club Hills, IL: International Code Council, Inc., 2012.

Resources

Books

The EcoNest Home, by Paula Baker-Laporte and Robert Laporte, New Society Publishers, 2015.

EcoNest: Creating Sustainable Sanctuaries of Clay, Straw and Timber, by Robert Laporte and Paula Baker-Laporte, Gibbs Smith, 2005.

Light Earth Building: Building with Wood and Earth, by Franz Volhard, Birkhauser, 2016.

Building Green: A Complete How-To Guide to Alternative Building Methods, by Clarke Snell and Tim Callahan, Lark Books, 2009.

Appropriate Building Materials, by Roland Stulz and Kiran Mukerji, Practical Action, 1993.

Building with Light-Clay: A Workbook, by Norbert Duerichen, 2012.

Earthen Floors: A Modern Approach to an Ancient Practice, by Sukita Reay Crimmel and James Thomson, New Society Publishers, 2014.

The Natural Plaster Book: Earth, Lime, and Gypsum Plasters for Natural Homes, by Cedar Rose Guelberth and Dan Chiras, New Society Publishers, 2002.

Clay Culture: Plasters, Paints and Preservations, by Carol Crews, Gourmet Adobe Press, 2010.

The Art of Natural Building: Design, Construction & Resources, by Joseph Kennedy, Michael Smith, and Catherine Wanek, New Society Publishers, 2009.

Web Resources

"Living Clay," by Perry A., Living Clay Co., livingclayco.com

"How to Build a Tumbler for Straw Light Clay (SLC)," wn.com

EcoNests:
Paula Baker-Laporte and Robert Laporte, econesthomes.com

Design Coalition:
designcoalition.org

The Last Straw journal:
thelaststraw.org (See, especially, thelaststraw.org/tribute-alfred-von-bachmayr. Bachmayr was the inventor of the tumbler.)

Day One Design, LLC (Oregon):

Natural home design and building services specializing in breathable walls and finishes, dayonedesign.org

Ion Ecobuilding:

A network of builders, designers, and suppliers focused on ecologically sustainable building. Publishes a directory of Pacific Northwest professionals, ionecobuilding.org

Communitecture:

A full service design firm for commercial and residential projects, communitecture.net

Index

Page numbers in *italics* indicate tables.

About the Author

Lydia Doleman is a licensed contractor with two decades of experience as a natural and sustainable builder, cabinet maker, and trainer. She has taught carpentry and natural building at Solar Energy International in Colorado and was lead ecological builder for Portland's City Repair project from 2002-09. Aiming to merge three converging passions in her life — art, ecology, and social justice — Lydia has created beautiful, high-performance, low-impact buildings across Oregon and Washington, including Portland's first permitted straw bale home, The Rebuilding Center's cob entryway, and a 3,300–sq. ft. light straw clay brewery. She's written articles for *The Last Straw Journal* and *Permaculture Activist* and appeared on NBC News and HGTV's *Off Beat America* on the topic of tiny homes, featuring a small-scale light straw clay timber frame home. Lydia lives in southern Oregon.

A Note About the Publisher

NEW SOCIETY PUBLISHERS is an activist, solutions-oriented publisher focused on publishing books for a world of change. Our books offer tips, tools, and insights from leading experts in sustainable building, homesteading, climate change, environment, conscientious commerce, renewable energy, and more — positive solutions for troubled times.

We're proud to hold to the highest environmental and social standards of any publisher in North America. This is why some of our books might cost a little more. We think it's worth it!

- We print all our books in North America, never overseas
- All our books are printed on **100% post-consumer recycled paper**, processed chlorine free, with low-VOC vegetable-based inks (since 2002)
- Our corporate structure is an innovative employee shareholder agreement, so we're one-third employee-owned (since 2015)
- We're carbon-neutral (since 2006)
- We're certified as a B Corporation (since 2016)

At New Society Publishers, we care deeply about *what* we publish — but also about *how* we do business.

New Society Publishers

ENVIRONMENTAL BENEFITS STATEMENT

For every 5,000 books printed, New Society saves the following resources:[1]

32	Trees
2,937	Pounds of Solid Waste
3,232	Gallons of Water
4,215	Kilowatt Hours of Electricity
5,339	Pounds of Greenhouse Gases
23	Pounds of HAPs, VOCs, and AOX Combined
8	Cubic Yards of Landfill Space

[1]Environmental benefits are calculated based on research done by the Environmental Defense Fund and other members of the Paper Task Force who study the environmental impacts of the paper industry.

Certified B Corporation

MIX Paper from responsible sources FSC® C016245 FSC www.fsc.org

new society PUBLISHERS www.newsociety.com